Proceedings of the First Ordinary Meeting
of the Regional Primatology Committee
for the Americas (CORP I)

PRIMATES OF THE AMERICAS

STRATEGIES FOR CONSERVATION AND SUSTAINED USE IN BIOMEDICAL RESEARCH

EDITED BY

Primo Arámbulo III
Filomeno Encarnación
Jaime Estupiñán
Hugo Samamé
Charles R. Watson
Richard E. Weller

Battelle Seattle Conference Center
Seattle, Washington, USA
October 29–31, 1990

SPONSORED BY

Pan American Health Organization
World Health Organization

BATTELLE PRESS
Columbus • Richland

Library of Congress Cataloging-in-Publication Data

Primates of the Americas: strategies for conservation and sustained use in biomedical research: proceedings of the first ordinary meeting, Regional Primatology Committee for the Americas (CORP-1), Battelle Seattle Conference Center, October 29-31, 1990/edited by Primo Arámbulo III ... [et al.]: sponsored by Pan American Health Organization.

p. cm.

English and Spanish.

Includes bibliographical references.

ISBN 0-935470-73-5: $34.95

1. Primates—Latin America—Congresses. 2. Primates as laboratory animals—Congresses. 3. Wildlife conservation—Latin America—Congresses. 4. Primates—Research—Latin America—Congresses. I. Arámbulo, Primo V., 1941- . II. Regional Primatology Committee for the Americas. III. Pan American Health Organization.

QL737.P9P685 1993 93-16987
639.9'798'098—dc20 CIP

Printed in the United States of America

Battelle Press
505 King Avenue
Columbus, Ohio 43201-2693
614-424-6393
1-800-451-3543

PREFACE

The first meeting of the Regional Primatology Committee for the Americas (known by its Spanish acronym: CORP) was convened by the Director of the Pan American Health Organization from 29 to 31 October 1990. Over 50 government representatives and scientists from 15 countries met at the Battelle Conference Center in Seattle, Washington, USA, to discuss the status of neotropical primates, and formulate plans for their preservation.

In the past decade, tremendous pressure has been brought to bear on the nonhuman primate populations indigenous to the Americas. This pressure is largely the result of habitat destruction through human encroachment and expansion of agricultural practices into former wilderness areas. Additional pressure has occurred from uncontrolled use of neotropical primates by the biomedical research community, which utilizes significant numbers in studies directed toward the solution of human health problems.

This situation has lead to rapid depletion and, in some cases, near extinction of many species of nonhuman primates unique to the ecosystems that they inhabit. In response to the uncontrolled exportation of their primates for legal and illegal purposes, source countries with indigenous populations subsequently banned or severely restricted the exportation of many species. However, even those actions have failed to achieve a satisfactory balance between the need for conservation and the need of the biomedical research community. The Regional Primatology Committee for the Americas was charged by the Pan American Health Organization with the responsibility of assessing the scope of the problem and developing recommendations and strategies for solving the problem. A comprehensive overview of all issues related to the rational protection and use of neotropical primates is beyond the scope of these proceedings. Rather, the goal of these proceedings is threefold: (1) to provide the reader with an overview concerning the status of and activities related to primate populations in the source countries; (2) to review policies and strategies intended to balance conservation and research needs; and (3) to present an integrated regional plan promoting habitat preservation and restoration, protection of threatened or endangered species, provision of key species for biomedical research, and promotion of programs intended to advance our knowledge of primate biology and medicine.

The establishment of the Regional Primatology Committee for the Americas was the culmination of efforts to develop a regional policy,based on the successful experience of the Peruvian Primatology Project as a model for the integration of issues related to conservation, reproduction in captivity and semi-captivity, and biomedical use of nonhuman primates. This policy was first expressed in the Declaration on Technical Cooperation among the Countries of the Americas in the Conservation and Utilization of Nonhuman Primates, known as the Declaration of Iquitos, that was the outcome of a gathering of some 50 scientists from Europe and the Americas at the Workshop on Controlled Breeding of Primates in Their Own Habitat, held at the Center for the Reproduction and Conservation of Nonhuman Primates, Iquitos, Peru, in November, 1986. The Declaration is the subject of a report presented to the Ministers of Agriculture of the Region at the V Inter-American Meeting, at the Ministerial Level, on Animal Health (RIMSA) in April 1987, during which the Ministers adopted Resolution VIII establishing the Regional Primatology Committee for the Americas.

This publication reviews the scientific issues upon which agreement was reached at the meeting, as well as the recommendations that were approved by consensus. Regional policy regarding the preservation of neotropical primates and their sustained use for the benefit of public health was a constant theme throughout the meeting. It is increasingly important in the changing world in which we live, inasmuch as it ensures the safeguarding of national patrimony.

Our sincere hope and intentions is that this publication, as a historical account, faithfully represents the spirit and content of the meeting.

William R. Wiley
Director
Battelle, Pacific Northwest Laboratories

Carlyle Guerra de Macedo
Director
Pan American Health Organization

PREFACIO

La primera reunión del Comité Regional de Primatología para las Américas (CORP) fue convocada por el Director de la Organización Panamericana de la Salud (OPS), del 29 al 31 de octubre de 1990. Más de 50 representantes de los Países Miembros de la OPS y científicos de 15 países se reunieron en el Centro do Conferencias de Battelle, en Seattle, Washington, Estados Unidos de América, para discutir la situación actual de los primates neotropicales y formular planes para su preservación.

En la década pasada las poblaciones de primates no humanos de las Américas experimentaron una tremenda presión, como resultado de la destrucción del habitat por las invasiones de la población humana y la expansión de las prácticas agrícolas en las áreas silvestres. En épocas pasadas, el uso no controlado de primates por parte de la comunidad de investigaciones biomédicas, causó una presión adicional, ya que se utiliza un número significativo de primates neotropicales en estudios dirigidos a la solución de problemas de salud humana.

Esto ha ocasionado una rápida disminución y en algunos casos la casi extinción de muchas especies de primates no humanos que son únicos en el ecosistema que ellos habitan. Como respuesta a la exportación sin control de sus primates para propósitos legales e ilegales, los países de origen de poblaciones autóctonas de primates neotropicales prohibieron y, de manera estricta, restringieron la transferencia de muchas especies. Sin embargo, esas acciones no han logrado establecer un balance satisfactorio entre la necesidad de conservación y las necesidades de la comunidad de investigaciones biomédicas. El Comité Regional de Primatología para las Américas fue creado por la Organización Panamericana de la Salud y se le encargó la responsabilidad de evaluar la magnitud del problema y de emitir recomendaciones y estrategias para su solución. Está fuera del alcance de estas memorias dar una visión resumida de todos los asuntos relacionados con la protección racional y el uso de los primates neotropicales; más bien, las metas de estas memorias son: 1) presentar al lector un resumen sobre la situación actual y las actividades en los países con población de primates neotropicales; 2) revisar las políticas y estrategias previstas para balancear las necesidades de conservación con las de investigación; y 3) presentar programas regionales integrados para promover la preservación y restauración del habitat, la protección de especies amenazadas o en peligro de extinción, la provisión de determinadas especies para investigación biomédica y la promoción de programas derigidos a avanzar en nuestro conocimiento de la biología y medicina de los primates.

El establecimiento del Comité Regional de Primatología para las Américas fue la culminación de los esfuerzos para desarrollar una política regional basada en la experiencia exitosa del Proyecto Peruano de Primatología, como un modelo para la integración de los asuntos relacionados con la conservación, reproducción en cautividad y semicautividad, asi como, el uso biomédico de primates no humanos. Esta política fue expresada por primera vez en la declaración sobre "La Cooperación Técnica entre los Países de las Américas en la Conservación y Utilización de los Primates No Humanos" conocida como la Declaración de Iquitos, que fue el fruto de medio centenar de científicos procedentes de Europa y de las Américas congregados en el Taller Sobre Crianza Controlada de Primates en su Propio Habitat, que se realizó en Iquitos, Perú, en noviembre de 1986. Esta Declaración fue objeto de un informe a los Señores Ministros de Agricultura de la Región en la V Reunión Interamericana sobre Salud Animal a Nivel Ministerial (RIMSA) realizada en abril de 1987. Conscientes de la demanda planteada en dicho informe los Señores Ministros aprobaron la

Resolución VIII por la cual se establece el Comité Regional de Primatología para las Américas.

Además, estas memorias tienen la particularidad que ofrecen los temas científicos que se trataron en esa reunión y que fueron objeto de controversias, esclarecimientos y asentimientos, dando génesis a las recomendaciones que emanaron de los participantes y que fueron aprobadas por consenso. La política continental que se refiere a la preservación de los primates no humanos y su uso sostenido en beneficio de la salud pública estuvo presente durante toda la Reunión. De la documentación que acompaña a estas memorias también fluye la conceptualización de esa política, que en este mundo cambiante deberá prevalecer con mayor vigor, ya que ella se constituye vigilante del patrimonio nacional.

Es la intención y esperanza sincera que esta publicación en fé representa el espíritu y contenido de la reunión. El sentimiento que ha guiado la preparación de estas memorias ha sido el deseo de contar con un recuento histórico de la reunión.

William R. Wiley
Director
Battelle, Pacific Northwest Laboratories

Carlyle Guerra de Macedo
Director
Organización Panamericana de la Salud

FOREWORD

It is a pleasure to be present at this meeting. Allow me to convey to you very cordial greetings from Dr. Carlyle Guerra de Macedo, Director of the Pan American Sanitary Bureau, who, due to prior commitments, is unable to be with us for the inauguration of the Regional Primatology Committee for the Americas (CORP).

The Director of the Office called this meeting in accordance with Resolution VIII of the V Inter-American Meeting on Animal Health at Ministerial Level (RIMSA), held in April 1987, establishing a CORP and requesting PAHO/WHO to be the Secretariat *ex officio*, and in keeping with Resolution XIV of RIMSA VI, of April 1989, which sets the guidelines for the composition of the Committee and ratifies that PAHO/WHO will act as Secretariat *ex officio*. It also makes a recommendation to create a fund for the conservation of nonhuman primates and to support the activities of this Committee.

These resolutions were ratified subsequently by the Governing Bodies of the Pan American Health Organization.

The policy stipulating conservation of the natural habitat and of its flora and fauna formulated by the countries of the Americas has been completely understood by the Governing Bodies of our Organization, which is the reason why, within our quadrennial technical cooperation programs, the conservation of nonhuman primates and their sustained use are accorded the hightest priority.

Nonhuman primates play an important role in public health. They are indispensable for the development of new vaccines and have contributed to the understanding of numerous diseases, such as diabetes, malaria, AIDS, yellow fever, measles, dengue, enteric infections, tuberculosis, mental disorders, and some forms of cancer.

Studies using nonhuman primates as models have yielded important findings for the development of diagnostic methods and vaccines for hepatitis and malaria, as well as in cardiovascular, genetic, dental, nutritional, oncogenic, ophthalmological, and behavioral research.

The availability of nonhuman primates for biomedical purposes is of worldwide concern and has different repercussions for the health services in each country. A great number of WHO priority programs require the sustained use of nonhuman primates, for example, in the potency testing and quality control of vaccines used by the Expanded Program on Immunization, the Program for Research on Tropical Diseases, the Special Program on Human Reproduction, etc.

The policies on conservation and sustained use of primates in public health are closely interrelated and must be developed in tandem so as to avoid any polarization that may cause antagonistic feelings, which in turn may have serious consequences. Despite the efforts and trends in developing and applying *in vitro* models in medical research, *in vivo* models will continue to be indispensable for some short and long-term research.

Today's world requires the development of a profound mutual understanding among conservationists and biomedical investigators, with a view to both keeping primate populations in their habitat and continuing to develop research that will contribute to the ongoing search for new knowledge in order to improve the quality of life.

Also, let us not forget that the programs aimed at the conservation of primates serve as vehicles for maintaining an ecological equilibrium between community development and the protection of the environment in neotropical jungles. The delivery of health services to the populations inhabiting the jungles of the Amazon Region is a clear example.

The Director of PAHO has entrusted me with the pleasant task of wishing you much success at the inauguration of the Regional Primatology Committee for the Americas and during its deliberations at this meeting.

Primo Arámbulo III
October 29, 1990

BIENVENIDA

Es muy grato estar presente en esta reunión. Les traigo un saludo muy cordial del Dr. Carlyle Guerra de Macedo, Director de la Oficina Sanitaria Panamericana, quién por compromisos previamente adquiridos no puede estar con nosotros para la instalación del Comité Regional de Primatología para las Américas (CORP).

El Director de la Oficina Convocó a esta Reunión en cumplimiento de la Resolución VIII de la V Reunión Interamericana de Salud Animal a Nivel MInisterial (RIMSA), que se realizo en abril de 1987, por la cual establece un CORP y solicita a la OPS/OMS ser la secretaría *ex-officio*, y también en atención a la Resolución XIV de la RIMSA VI, de abril de 1989, por la cual establece los lineamentos de composición del Comité y ratifica que la OPS/OMS actue como Secretaría *ex-officio*. Además recomienda la creación de un fondo para la conservación de los primates no humanos y apoyar las actividades del presente Comité.

Estas resoluciones fueron ratificadas posteriormente por los Cuerpos Directivos de la Organización Panamericana de la Salud.

La política conservacionista de los habitat natural y de su flora y fauna por parte de los países de las Américas ha sido muy bien comprendida por los Cuerpos Directivos de nuestra Organización, razón por la cual dentro de nuestros programas cuadrienales de cooperación técnica, la conservación de los primates no humanos y su uso sostenido se ubica en la mas alta prioridad.

Los primates no humanos juegan un papel importante en la salud pública. Ellos son imprescindibles en el desarrollo de nuevas vacunas y han contribuido en el conocimiento de numerosas enfermedades como la diabetes, la malaria, el SIDA, la fiebre amarilla, el sarampión, el dengue, las infecciones entéricas, la tuberculosis, los trastornos mentales y algunas formas de cáncer.

Estudios en primates no humanos como modelos dieron allazgos importantes en el desarrollo de métodos de diagnóstico y vacunas contra la heptatitis y la malaria, así como en investigaciones cardiovasculares, genéticas, dentales, nutricionales, oncogénicas, oftalmológicas y de comportamiento.

La disponibilidad de primates no humanos para fines biomédicos es una preocupación mundial, que tiene repercusiones distintas para los servicios de salud de todos los países. Un gran número de programas prioritarios de la OMS requieren el uso sostenido de primates no humanos. Por ejemplo, en la prueba de potencia y control de calidad de las vacunas que utiliza el Programa ampliado de inmunizaciones, en el Programa de investigación sobre enfermedades tropicales, y el Programa especial de reproducción humana, entre otros.

Las políticas de conservación y uso sostenido de primates en salud pública están íntimamente ligadas, debiendo ser desarrolladas paralelamente con el fin de evitar la polarización que puede ocasionar sentimientos antagónicos con graves consecuencias. A pesar de los esfuerzos y tendencia para desarrollar y aplicar modelos *in vitro* en investigación médica, la necesidad de modelos *en vivo* seguirá siendo indispensable en algunas investigaciones a corto y largo plazo.

El mundo actual requiere una íntima compresión entre los conservacionistas y los investigadores biomédicos, con el propósito de mantener las poblaciones de primates en su habitat, por un lado, y por el otro seguir desarrollando las investigaciones que contribuyen a la constante búsqueda de nuevos conocimientos para mejorar la calidad de vida.

No debemos olvidar, además que los programas de conservación de primates sirven como vehículos para mantener el equilibrio ecológico entre desarrollo comunitario y la protección del medio ambiente en las selvas neotropicales. La prestación de servicios de salud a las poblaciones selvícolas en la amazonia es un claro ejemplo.

El Director de la OPS me dio el grato encargo de desearles el mayor de los éxitos en la instalación del COmité Regional de Primatología para las Américas, y en sus deliberaciones en esta reunión.

Primo Arámbulo III
29 de octubre de 1990

ACKNOWLEDGEMENTS

The editors wish to express their sincere gratitude to the CORP I members, participants, observers and secretariat whose contribution, active participation, and willingness to share their experiences were crucial for the achievement of the objectives of the Meeting and for the production of this publication.

Special thanks are given to the World Health Organization Collaborative Center for Training and Reference in Clinical Pathology of Neotropical Primates, Battelle Pacific Northwest Laboratories for the support provided which contributed to the success of the meeting. The success of the meeting was also aided greatly by several individuals who devoted a significant amount of time to its planning and coordination. We are particularly grateful to the special efforts of Ms. Toni Quackenbush of Battelle Pacific Northwest Laboratories and Ms. Ana Maria C. Flores of the Veterinary Public Health Program of PAHO/WHO.

This document was assembled electronically using WordPerfect software on a personal computer. The texts of the various papers were received from the authors on diskette and converted to a standard format. Translations were provided by the staff at PAHO. The body text is printed in GC Times fonts, and appendices in Helv. The camera-ready copy was printed on a LaserJet III[1].

Richard Weller
Charles Watson
September, 1992

[1] Trademark Notification: WordPerfect is a registered trademark of WordPerfect Corporation, Orem, Utah; GC Times is a product of Agfa Corporation, AGFA Compugraphic Division; Helv is a product of IQ Engineering; LaserJet is a registered trademark of Hewlett-Packard Company

AGRADECIMIENTOS

Los editores desean expresar su sincera gratitud a los Miembros del CORPI, participantes, observadores y secretaría, cuya contribución, participación activa, y voluntad de compartir experiencias fue crucial para el logro de los objetivos de la Reunión y para la producción de esta publicación.

Agradecimiento especiales también se presentan al Centro Colaborador de la Organización Mundial de la Salud, sobre Entrenamiento y Referencia en Patología de Primates Neotropicales, de los Laboratorios Battelle Pacific Northwest, por el apoyo brindado y que contribuyo al éxito de la Reunión. Para el éxito de la Reunión también se contó con colaboración de varias personas que dedicaron tiempo. De manera particular se agradecen los esfuerzos de la Señora Quackenbush de los Laboratorios Battelle Pacific Northwest y la Señora Ana María C. Flores del Programa de Salud Pública Veterinaria de la OPS/OMS:

Este documento fue procesado electrónicamente utilizando el programa Word Perfect en un computador personal. Los textos de varios trabajos fueron recibidos de los autores en diskette y convertidos en un formato standard. La traducción fue hecha por los técnicos de la OPS. El texto se imprimo con las fuentes GC times, y los apéndices en fuentes Helv. La copia final se imprimió en un impresora LaserJet III[2].

Richard Weller
Charles Watson
septiembre, 1992

---

[2] WordPerfect esta una marka registrada de WordPerfect Corporation, Orem, Utah; GC Times es un producto de Agfa Corporation, AGFA Compugraphic Division; Helv es un producto de IQ Engineering; LaserJet esta una marka registrada de Hewlett-Packard Company

TABLE OF CONTENTS

CONTENIDO

Section 2 - in SPANISH / Sección 2 - en ESPAÑOL

Appendices / Anexos

PRIMATES OF THE AMERICAS

STRATEGIES FOR CONSERVATION AND SUSTAINED USE IN BIOMEDICAL RESEARCH

Section 1

ENGLISH

Sección 1

INGLES

FINAL REPORT

I Ordinary Meeting of the Regional Primatology Committee for the Americas

(CORP 1)

Seattle, Washington, USA

29-31 October 1990

FINAL REPORT

The I Meeting of the Regional Committee on Primatology for the Americas was held in the Seattle Conference Center, in the State of Washington, United States of America, in accordance with the convocation made by the Director of the Panameican Health Office, in compliance with Resolutions VIII and XIV of the V and VI Interameican Ministerial Level Meetings on Animal Health, and subsequently ratified by the Governing Bodies of the Panamerican Health Organization (OPS/OMS).

GOVERNING BOARD

By unanimous decision, the Governing Board of the Meeting was comprised as follows:

Chairman:	Ing. Manuel Uceda General Director of Forestal and Fauna Ministry of Agriculture Peru
Vice Chairman:	Dr. Stephen Potkay Director Veterinary Resources Program National Institutes of Health United States
Members:	Dr. Alejandro Estrada Tropical Biology Station National Autonomous University of Mexico Mexico
	Dr. Jordan Paulo Wallauer Deparment Head for Wild Fauna and Flora National Secretariat of the Environment Brazil

Dr. Primo Arámbulo III, Coordinator of the Public Veterinary Health Program (OPS/OMS), representing the Director of the Panamerican Health Organization, Dr. Carlyle Guerra de Macedo, served as Secretary _ex-officio_ and Reporter.

PARTICIPANTS

The following governments, charter members of the Committee, were represented at the Meeting: Argentina, Barbados, Bolivia, Brazil, Canada, Colombia, United States, Mexico, Paraguay, and Peru.

Also in attendance as observers were representatives of institutions from the following countries: Brazil, Colombia, United States, France, Guatemala, Japan, and the Netherlands, as well as the International Council for Laboratory Animal Science (ICLAS).

The Battelle Pacific Northwest Laboratories, served as the host institution.

The list of participants is attached as Annex 1

SCHEDULE OF TOPICS AND SCHEDULE OF SESSIONS

The Meeting consisted of an inaugural session, three plenary sessions, and a closing session.

In the first plenary session, restricted to the charter members of CORP, the Governing Board was elected, and a committee, comprised of the delegates from Argentina, Colombia, and Paraguay, was named to study the proposed CORP Charter and By-Laws.

In this same session, three other committees were established. The first, on the mobilization of resources, under the direction of Dr. James Wong of Canada; the second, on primatology research, under the direction of Dr. Julio C. Ruiz of Argentina; and the third, on the conservation of non-human primates, under the direction of Dr. Jose Vicente Rodriguez of Colombia.

The schedule of sessions was subsequently approved (see Annex 2).

In the inaugural session, at the request of the Chairman, Dr. Roger anderson of Battelle Pacific Northwest Laboratories, who on behalf of said institution, did express their satisfaction in collaborating in the realization of the First Meeting of CORP, and at the same time stated that Battelle is aware of the importance of primates to the sciences.

Dr. Primo Arambulo III, representing the Director of the Panamerican Health Organization, offered a very special greeting to the delegations and observers participating in the meeting, stated the objectives of the convocation, and outlined the policy of the OPS/OMS with respect to the conservation and sustainable use of non-human primates.

At the request of the Chairman, Dr. Jordan Wallauer took the floor and on behalf of the Special Secretariat of the Environment of Brazil, did express gratification for the realization of this initiative. Subsequently, the Chairman asked Dr. Hernan Zeballos, Undersecretary of Agricultural Development of Bolivia, to open the Meeting. The undersecretary stated that the participation at this I Meeting of the Committee will permit the identification of

alternatives to control the destruction of the environment, which is associated with the impoverished constitutions of the populations situated in the forested regions, and votes were made for the I Meeting of CORP to achieve the proposed objectives.

In the second plenary session on conservation management, and health, the following topics were addressed: "Current Status and Conservation of Neotropical Primates in their Native Habitat", presented by Dr. Roderick Mast of Conservation International; "The Status of Natural Primate Populations and the Benefits of Sustained Cropping", by Biologist Filomeno Encarnacion, temporary Advisor to OPS/OMS; "Environmental Factors for the Breeding of Primates in Special Habitats", by Dr. Thomas Wolfle of the Academy of Science; "the Conservation of Non-Human Primates and their Importance to Publica Health", by Dr. Stephen Potkay, Director of Veterinary Resources Services of the National Institutes of Health of the United States; and "The Technical Cooperation Program of the Panamerican health Organization", by Dr. Primo Arambulo III, Coordinator of the Veterinary Public Health Program of the Panamerican Health Organization (OPS/OMS).

The Chairman offered the floor to the delegates from Argentina, Barbados, and Paraguay, who reported on the status of primatology in their respective countries.

In the third plenary session, topics relating to politics and economics were addressed. The topic "Management of Fauna in Brazil" was presented by Dr. Jordan Paulo Wallauer, of the Special Secretariat of the Environment of Brazil; "Management of Non-Human Primates as a Renewable Resource: The Economics of Conservation", by Dr. Jose Vicente Rodriguez, of the Natural Resources Institute of Colombia; "Multisectoral and Intercountry Cooperation for the Development of National Primatology Programs", by Ing. Manuel Uceda, General Director of Forestry and Fauna of the Ministry of Agriculture of Peru; "The Mobilization of Resources for the Conservation of Ecosystems", by Dr. Edward Liebow, of Battelle Pacific Northwest Laboratories; "Community Participation in the Protection of the Natural Habitat of Neotropical Primates and the Natural Resources Policies of Bolivia", by Dr. Hernan Zeballos, Undersecretary of Agricultural Affairs for the Ministry of Rural and Agricultural Affairs of Bolivia; and "The Regional Action Plan of Primatology for the Americas", by Biologist Filomeno Encarnacion, Temporary Advisor to OPS/OMS.

Subsequently, the CORP Charter and By-Laws committee made proposals to the Assembly for the changes to the reviewed drafts, which were approved. The charter and By-Laws are part of this Final Report as Annex 3, and later by unanimous decision the assembly approved the general objectives and the strategies for the Regional Primatology Plan of Action for the Americas.

The Chairman made a request to the committees on Research, Conservation and Resource Mobilization to make a report of their recommendations which were approved.

The assembly unanimously approved a vote of applause for Battelle Pacific Northwest Laboratories for the excellent facilities and hospitality provided for this meeting.

In the closing session the chairman of the Meeting, Ing. Manuel Uceda, and Dr. Primo Arambulo III of the OPS/OMS took the floor to speak.

APPROVED RECOMMENDATIONS

The I meeting of the Regional Committee approved the following recommendations:

TO THE COUNTRIES:

1. To request the developed countries to channel economic resources toward countries that have native nonhuman primate populations so that they can develop programs for the conservation and sustainable use of primates for their biomedical research.

2. To urge those countries that have not already done so to establish and/or strengthen their national projects for the conservation of nonhuman primates in their natural habitat, in captivity, or in semicaptivity.

3. To call upon the countries to establish bilateral agreements for the development of joint activities for the conservation of natural resources, with emphasis on nonhuman primate populations.

4. To call upon the countries, insofar as possible, to adopt strategies and actions from the Regional Plan of Action in Primatology for the Americas in their national plans and programs.

TO THE PAN AMERICAN HEALTH ORGANIZATION (PAHO/WHO):

1. To request PAHO/WHO to facilitate and coordinate the transfer of primatology-related technology, especially in the member countries of CORP.
2. To request PAHO/WHO to formulate technical cooperation activities based on the overall objective and strategies contained in the Regional Plan of Action in Primatology for the Americas, approved at this I Meeting of CORP.

3. To call upon PAHO/WHO to continue cooperating in the search for economic and technical resources for the development of primatology in the countries of the Region.

4. To call upon PAHO/WHO to serve as secretariat ex officio of the intercountry agreements referred to in Recommendation 3 to the countries.

5. To request PAHO/WHO to ask the Pan American Health and Education Foundation (PAHEF) to set up a project heading for the deposit of grants and donations from individuals and/or public and private institutions, among others, for the development of primatology and the activities of CORP.

Recommendations of the Committees on Research, Conservation, and Resource Mobilization

Committee on Research

1. To stimulate research on optimizing the breeding of platyrrhine monkeys in captivity in the neotropical countries.

2. To stimulate research on the natural history of neotropical primates with a view to promoting their conservation in forest settings.

3. To train human resources in basic and applied research in the neotropical countries.

4. To establish, in each institution or university engaged in the study of primatology, a documentation and/or reference center offering up-to-date literature and links with databases that contain biological information, especially in the area of primatology.

This committee was chaired by Dr. Julio C. Ruiz of Argentina, and its members were Dr. María T. Muñoz of Paraguay, Dr. Milton T. de Mello of Brazil, and Dr. Alejandro Estrada of Mexico.

Committee on Conservation

1. To ensure that CORP is involved in development of the Plan of Action for the Neotropics proposed by the primate specialists group at UICN.
2. To channel resources from institutions that use primates in biomedical research toward to the identification of new conservation areas and/or implementation of those that already exist.

3. **To condition the use of nonhuman primates, starting with the production of individuals under natural or controlled conditions, on the investment of economic resources comparable to the real cost of protecting them and to prices on the international market.**

4. **To incorporate the objectives of the national primatology programs in bi- and multinational agreements such as the Amazon Cooperation Treaty.**

5. **To initiate short- and long-term training on different aspects related to the management of primates in natural and/or controlled environments.**

6. To promote knowledge, using proactive methods, about the diversity of primates in each neotropical region, including information on each of the species and messages about their importance and conservation.

7. To develop a database on conservation status and management, through organizations of recognized international prestige, that will include, in addition, information on primate-raising in the countries of origin and of destination.

8. To encourage the standardization of systematic criteria that will make it possible to standardize knowledge about the presence and distribution of species.

This committee was chaired by Dr. José Vicente Rodríguez of Colombia, and its members were Dr. Jordán Paulo Wallauer, Dr. Alejandro Estrada of Mexico, the biologist Filomeno Encarnación of Peru, and Dr. Roderick Mast of Conservation International.

Commission on Resource Mobilization:

The member countries are responsible for submitting a document to the secretariat by 31 April 1991 that defines their sources and identifies their objectives.

In terms of definition of sources, the member countries have been divided into those that have native primate species, e.g. Peru, and those that do not, e.g. the United States of America. The following outline will be useful in completing the document:

Countries with native primate populations:

1) Number of species and population estimate for each species.

2) Identification of any existing primate center and/or other institution(s) involved in primate conservation and production.

3) Description of the facilities, including personnel, duration of financing, affiliation(s), and other pertinent data.

4) Government agencies responsible for authorizing the export of nonhuman primates.

Countries without native primate populations:

1) Skill and experience in raising nonhuman primate colonies in captivity, health and diagnostic services. Organization of workshops for transfer of information to the member countries.

2) List of investigators that need nonhuman primates in the countries of origin or for export to research institutions.

3) Provision of information to the scientific community on the availability of nonhuman primates.

4) List of institutions that have or use neotropical primates in research, diagnostic testing, education, and zoological parks.

5) List of requirements for every neotropical species used in these programs.

The objectives are the same for all the member countries. Accordingly, each one should specify the strategies by means of which these objectives will be achieved.

1. Exchange of information and collaboration on research projects among the member countries.

2. Development of infrastructure to ensure the conservation of species and the maintenance of research potential in the countries of origin.

3. Preservation and economic promotion of the primate sources in the countries of origin through the establishment environmental education and conservation programs.

The commission is chaired by Dr. James Wong of Canada and the members are Mr. Jean Baulu of Barbados and Dr. Stephen Potkay of the United States of America.

IN WITNESS WHEREOF, the Chairman of the Meeting and the Secretary ex oficio do sign this Final Report in the Spanish and English languages, whose texts will have equal authentic value.

DONE in the city of Seattle, State of Washington, United States of America on the thirty-first day of October of nineteen hundred ninety. The Secretary will deposit the original texts in the archives of the Panamerican Health Office, and will send copies to the Members of the Committee and to the Member Governments of the Organization.

Ing. Manuel Uceda
Director General de Forestal y Fauna
Ministerio de Agricultura del Peru
Presidente

Dr. Primo Arambulo III
Secretario ex-officio
del Comité Regional de
Primatología para las Américas
Oficina Sanitaria Panamericana

NEOTROPICAL PRIMATE CONSERVATION

Russell A. Mittermeier
Department of Anatomical Sciences HSC,
State University of New York,
Stony Brook, NY 11794, U.S.A.
and
Conservation International, 1015 18th
Street, NW Washington, DC 20036, USA

Warren G. Kinzey
Department of Anthropology,
City College and the Graduate Center,
City University of New York,
Convent Avenue, E. 138th Street,
New York, NY 20031, U.S.A.

Roderic B. Mast
Conservation International,
1015 18th Street, NW,
Washington, DC 20036, U.S.A.

The greatest threat to Neotropical primates is the loss of tropical forest for agriculture, ranching, commercial logging and hydroelectric projects. Hunting of primates for consumption or sale and live capture for the pet trade and for biomedical purposes is also a threat to certain species. Twenty-eight of the 76 Neotropical primate species are considered endangered and the highest density of endangered taxa occurs in Brazil's Atlantic forest, followed by Amazonia and Middle America. The most endangered Neotropical primate genera are *Brachyteles* and *Leontophithecus*, followed by some species of *Ateles, Lagothrix, Saguinus* and the Central American squirrel monkey (*Saimiri oerstedii citrinellus*). Solutions to the problems of primate conservation are discussed, including habitat conservation, protection against poaching and international primate trade, education, research and captive propagation, and examples of successful conservation efforts are provided.

THE PROBLEMS

Major threats to wild primates throughout the world can be categorized as either direct pressures against the animals themselves, or destruction and alteration of their habitat. The relative magnitude of these two hazards varies among species and regions, but one or both of them influence virtually all existing primate populations.

THREATS TO HABITAT

The most significant factor contributing to the decline of the world's non-human primates is habitat destruction (Mittermeier and Cheney 1987) arising largely from the economic development of natural environments (Myers 1976). More than 90% of all primate species occur in tropical forests and as these forests disappear, so do the animals that depend on them for survival.

The ultimate cause of habitat destruction is the human population explosion, which is greatest in the poor, underdeveloped countries of the tropics. The addition of millions of landless people each year to the already overburdened developing countries makes the trend toward large-scale forest loss seem irreversible (Mittermeier and Cheney 1987). Population growth rates in almost all tropical countries are far higher than in the developed world. The annual population growth rate in the

United States is 0.86% and rates actually fall below zero in some of the European nations; in sharp contrast, Brazil's annual growth rate is 2.07%, Indonesia's is 1.74%, Zaire's is 3.04% and Peru's is 2.51% (International Institute for Environment and Development and World Resources Institute, 1987), and these are among the world's top countries for primary diversity and tropical forest cover (Mittermeier 1987d). Though it is inevitable that a portion of the world's forests and primates will disappear, the role of conservationists must be to limit this loss as much as possible.

Neotropical forests are being destroyed for a variety of reasons: agriculture, ranching, demand for charcoal and wood for fuel, commercial logging of hardwoods, road building and massive hydroelectric projects. Recent estimates, although differing somewhat, are in general agreement that a minimum of 0.6%/year of all closed, tropical broad-leaved forests were being cut in the late 1970s (Lanley 1982; Melillo et al. 1985), and this figure is undoubtedly higher now, given the steady increase in development activities that has taken place in the tropics.

The extent and effects of habitat destruction vary from one region to another, although estimates are rough. In Bolivia, only about 0.2% of the forest is being destroyed every year, but in the Atlantic forests of Brazil, so much has already been destroyed that any further loss of forest constitutes a grave threat to the remaining primate populations, many of which are already endangered. Similarly, if present trends continue, many of the forests of Central America will also disappear by the early 1990s (Myers 1984).

The rapid rate of forest destruction in some areas emphasizes the vulnerability of those regions that still include huge tracts of forest. For example, since the advent of Europeans, the Atlantic forests of eastern Brazil have been reduced to between 1 to 5% of their original size. Since most of this destruction has occurred in the last 20 or 30 years, it is easy to imagine how vast expanses of forest such as those found in Amazonia could also be vulnerable to rapid devastation (Mittermeier et al. 1982). If present trends continue, it is estimated that by the year 2020 the forests of Amazonia will have been reduced by 50%, and that all other accessible forests outside the boundaries of parks and reserves will have been completely destroyed (Myers 1987). Other estimates indicate that destruction in Amazonia could be even more rapid with most of the forest gone by the end of the century or earlier (Fearnside 1982).

Agriculture

The single most serious cause of forest destruction worldwide is conversion for agriculture, either from traditional slash-and-burn subsistence farming or large-scale commercial cultivation. In fertile areas such as eastern Brazil, most forests have already been replaced by permanent farms or plantations (Mittermeier et al. 1982). In areas of low soil fertility (e.g., much of Amazonia), the soil can support only a few seasons of intensive farming before erosion and overuse deplete its nutrients. If cleared areas are small, and followed by a fallow period, such farming can be ecologically sound, however, the rate of population growth far exceeds the low yield of these careful approaches. The result is that degraded plots are abandoned and new areas of forest are cleared to begin anew the cycle of destruction (Department of State 1980).

Cattle Ranching

The second major cause of forest destruction in Central and South America is cattle ranching. In contrast to subsistence farming, cattle ranching often involves foreign-owned corporations and active government support. Brazil, for example, encourages cattle ranching in Amazonia as part of its campaign to extend control over the interior, and to develop its export market. Large ranching consortiums in Brazil include such multi-national corporations as Volkswagen Corporation, which holds a concession for 1400 square km of Amazonian forest (Myers 1984). Ironically, cattle ranching in tropical forest areas is much less efficient than on existing sub-tropical pastures, but the terms offered by the government to encourage the development of Amazonia are very attractive; there is therefore very little inducement to adopt better land management practices (Myers 1984).

Trees for Fuel and Charcoal

The cutting down of trees for fuel is also a significant threat which accounts for about 80% of all wood removed from tropical forests. The need for firewood and charcoal is usually thought to have only a local impact, but there is a large international trade in charcoal as well as it is causing a major worldwide depletion of tropical forests (Office of Technological Assessment, 1984). In some countries, fuel use is the major cause of deforestation [e.g., parts of Central America (Myers 1984)]. At least 1.5 billion people living in the Third World are thought to be affected by a scarcity of fuelwood, and forests are suffering from increasing harvesting rates (Myers 1984). Throughout eastern Brazil, the kilns in which native trees and introduced eucalyptus are converted to charcoal are widespread.

Commercial Logging

Industrial logging, both for local use and exportation, accounts for about 20% of the total volume of wood removed from tropical forests. Such logging depletes the resources of local people while, in the case of exportation, the major benefits go to foreign consumers, national governments or large companies. Japan is particularly dependent on foreign sources of pulp, and relies increasingly on South American hardwood trees (Myers 1984). Although commercial extraction of wood is generally selective, it still has an adverse effect on primate species such as *Ateles paniscus* and *Lagothrix lagothricha* that depend on mature primary forest (Wolfheim 1983; Skorupa 1986). Moreover, there is growing evidence that the productivity of the residual trees is severely reduced by selective logging (Skorupa 1986). Sadly, although companies are usually only granted licenses to remove small portions of a forest, local corruption and/or lack of enforcement frequently results in abuse of contracts. Finally, logging roads constructed to remove timber from the forest often do more damage than tree removal itself.

Hydroelectric Projects

Hydroelectric projects often destroy some of the least disturbed and most important Neotropical forest areas, including those in parks or reserves. Such "protected" areas are often chosen as project sites, both because they are relatively uninhabited and because they are already in government hands. In some cases, the size of the inundated area can be quite large. The Tucurui

Project in Amazonian Brazil, for example, flooded approximately 3000 square km (A. Johns, personal communication) and other projects of comparable size are planned for the region as well.

DIRECT THREATS TO ANIMALS

Hunting

Primates are hunted for a variety of reasons, but most significantly as a source of food. As with habitat destruction, the effects of hunting vary considerably from region to region and among species. Hunting of primates for food poses a major threat in the Amazonian region, where thousands of primates are killed each year for immediate consumption or sale in local markets. In some places (e.g., the interior of Suriname; Amazonian Ecuador), as much as 35% of the meat eaten by the forest people is derived from primates (Mittermeier 1977, 1987b; Yost and Kelley 1983). Hunting of primates is often prohibited by law, but enforcement is rare or nonexistent in the remote areas where most of it occurs (Mittermeier and Coimbra-Filho 1977; Mittermeier 1987b).

When hunting is common it does not affect all species equally. Amazonian hunters prefer large-bodied, prehensile-tailed monkey species (Peres 1987) and the woolly monkeys, spider monkeys and, to a lesser extent, howlers and capuchins are heavily preyed upon in most parts of Amazonia. These species have become locally extinct in many parts of their former ranges, even in otherwise suitable primary forest habitats (Mittermeier and Coimbra-Filho 1977; Soini 1982; Mittermeier 1987b). Smaller primates, such as squirrel monkeys and tamarins, are rarely persecuted, as they barely recompense the hunter for his shotgun shell.

Primates are also killed to obtain skins or other body parts for ornaments. In Amazonia, for example, various tribes use monkey teeth, skulls and skins for ornamentation and these items occasionally find their way into tourist shops. The bushy tails of saki monkeys (genus *Pithecia*) are popular as dusters, and monkey skins, monkey skin caps and stuffed monkeys are frequently seen in the curio shops of larger Amazonian cities. By and large, however, these type of primate utilization are a by-product of hunting for food.

Finally, some Neotropical primates are hunted as agricultural pests. Although semi-terrestrial Old World species such as baboons, macaques and vervets are worse offenders, Neotropical primates, especially *Cebus*, are crop pests in many parts of South America. Indeed, the vernacular names for *Cebus nigrivittatus* in Suriname (*nyangkaru mongi*) and for *Cebus apella* in Colombia (*maicero*) translate to "corn eater" (Mittermeier 1987b). Killing of primates as alleged crop pests may result in losses greater than expected. As primate habitats continue to be encroached upon, the more adaptable species will probably rely more heavily on crops. The unfortunate result will be heightened conflict between human and non-human primates which, in turn, will exacerbate the difficulty of gaining real support for conservation.

Live Capture

The second direct threat to primates is live capture for the pet trade and biomedical research. At the height of the international primate trade, during the 1950s and 1960s, the exportation of primates from Central and South America for these purposes involved hundreds of thousands of animals (Mack and Mittermeier 1984). Species that have been hurt by international trade include the squirrel monkey, exported from Colombia and Peru for the pet trade; cotton-top tamarins from northern Colombia, trapped for research labs (Hernandez-Camacho and Cooper 1976); woolly monkeys and spider monkeys, which are still popular as pets in Amazonia (Mittermeier and Coimbra-Filho 1977) and the golden-headed lion tamarin, a highly endangered form that has recently entered the illegal trade both within Brazil and internationally (Mallinson 1984).

The local trade in primates within source countries is also potentially significant. Some developing countries, such as Brazil, are beginning to use indigenous species in their own biomedical research programs and hundreds of marmosets (*Callithrix* spp.) are sold each year in illegal pet markets in the large cities of Rio de Janeiro and Sao Paulo. Many of these primates find their way to the international market through pet dealers in countries such as Japan.

In Amazonia, where primates are already hunted for food, their role as household pets is a by-product of meat hunting. Infant primates usually make the best pets and are often obtained by shooting an adult female and removing its clinging baby. For primates such as *Lagothrix lagothricha* even the small added incentive of the pet trade may add significantly to the conservation problems. Since woolly monkeys are both the favorite primate meat source and the most desirable primate pets, a hunter will usually shoot females with infants in preference to other animals. The effects of such selective hunting on the reproductive rates of a slow-breeding species such as the woolly monkey clearly can be dangerous.

THE MOST ENDANGERED NEOTROPICAL PRIMATES

Of the 76 species of New World monkeys currently recognized, 28 species (and/or subspecies) are already listed as endangered, vulnerable, rare or indeterminate in the IUCN Red Data Book (Thornback and Jenkins 1982; see Table 1). This represents only a minimum estimate of the number of endangered Neotropical primates, as there are many species with unknown conservation status.

When one considers the distribution of endangered taxa by region, the highest density (two genera and nine species) occurs in the Atlantic forests of Brazil, followed by Amazonia and Middle America (see Mittermeier 1987d; Figure 4). In terms of countries, Brazil, with 19 endangered taxa (18 of them endemic), is also at the top of the list.

The most critically endangered New World monkeys are the genera *Brachyteles* and *Leontophithecus* (Table N° 1). Both occur only in small remnant patches of the Atlantic forest region in eastern Brazil, and both are on the verge of extinction. The muriqui (*Brachyteles arachnoides*) is the largest South American primate. Its remaining population may include as few as 300 to 400 individuals (Mittermeier 1987b,c). Similarly, the lion tamarin genus, *Leontopithecus* includes three species (Rosenberger and Coimbra-Filho 1984) that are all highly endangered. They occur in only a few remnant forest patches, and the reserves created for them are in some cases small and partially occupied by colonists (Mittermeier et al. 1982). The threats to both genera are discussed in greater detail in Mittermeier (1987a).

TABLE N° 1. The Most Endangered Neotropical Primate Taxa

Endangered genera
Brachyteles
Brachyteles arachnoides
Leontopithecus
Leontopithecus rosalia
Leontopithecus chrysomelas
Leontopithecus chrysopygus

Endangered Species	Endangered Subspecies
Callithrix aurita	*Callithrix argentata leucippe*
Callithrix flariceps	
Callithrix geoffroyi	*Saguinus bicolor bicolor*
Saguinus oedipus	*Chiropotes santanas satanas*
Callicebus personatus	*Cacajao calrus calrus*
C. p. personatus	
C. p. nigrifrons	*Cebus apella xanthosternos*
C. p. melanochir	*Alouatta belzebul ululata*
Saimiri oerstedii	*Ateles geoffroyi azuerensis*
S. o. oerstedii	*Ateles geoffroyi panamensis*
S. o. citrinellus	*Ateles geoffroyi grisescens*
Alouatta fusca	*Ateles fusciceps fusciceps*
A. f. fusca	
A. f. clamitans	*Ateles belzebuth hybridus*
	Ateles belzebuth marginatus
Lagothrix flavicauda	

Other seriously endangered New World primates are the spider monkeys (*Ateles*) and the woolly monkeys (*Lagothrix*). Although both have large geographical ranges, they are heavily hunted and in most cases require relatively undisturbed forest for their survival. Moreover, because of their large size, they reproduce at relatively slow rates (in *Ateles*, inter-birth intervals are about three years [Eisenberg 1979]) and therefore recover slowly from any kind of exploitation. Not surprisingly, *Ateles* and *Lagothrix* have disappeared from many parts of their former distribution, even in areas of suitable primary forest (Mittermeier and Coimbra-Filho 1977; Mittermeier 1987b).

Although the majority of endangered Neotropical primate species occur in the Atlantic forest and Amazonia, endangered primates endemic to other regions include the yellow-tailed woolly monkey (*Lagothrix flavicauda*) of northern Peru, the Central American squirrel monkey (*Saimiri oerstedii* and the Panamanian tamarin (*Saguinus oedipus geoffroyi*). The yellow-tailed woolly monkey, the only Andean primate species, is restricted to a small area in northern Peru. Rediscovered in 1974 after a lapse of almost 50 years (Mittermeier et al. 1975, 1977), it is threatened by habitat destruction and hunting in its once isolated cloud forest range.

The Central American squirrel monkey is found only in western Panama and adjacent parts of southeastern Costa Rica. Although squirrel monkeys are usually among the most abundant Neotropical primates, *S. oerstedii* is considered highly endangered. A recent ecological study and population survey of this monkey in Costa Rica revealed that there are no more than 3000 individuals remaining (Boinski 1986) and it is possible that these Costa Rican animals are all that still survive. Fortunately, each of the two subspecies is represented in one of Costa Rica's national parks, *S. oerstedii* at Corcovado National Park, and *S. o. citrinellus* at Manuel Antonio National Park, though in the latter case, population estimates are low and the habitat is severely degraded.

The cotton-top tamarin is another highly endangered monkey. It is found only in northwestern Colombia, an area that has been subjected to heavy forest destruction during the past two decades. Many individuals were also collected for the primate trade in the 1960s and early 1970s. It is estimated that some 20,000 to 30,000 were exported during the peak years of the trade (Hernandez-Camacho and Cooper 1976). *S. oedipus* is well represented in captivity, with some 1161 in zoos and research labs in the United States and Europe. A studybook has recently been completed for this species (Tardiff and Colley 1987).

Little up-to-date information exists on the status of the cotton-top tamarin in the wild. However, it was already depleted when the last survey took place in 1975 (Neyman 1977). Preliminary observations from a recent survey indicate that the situation has deteriorated since then (V. Rodriguez and J. Hernandez-Camacho, personal communication). It is likely that cotton-tops are in a position approaching that of the lion tamarins in Brazil.

POTENTIAL SOLUTIONS

Potential solutions to the problems of primate conservation in the Neotropics (Figures N° 1 and 2) involve habitat conservation, protection against poaching and uncontrolled live capture, education and public awareness, training (particularly of local conservationists), research on wild populations, captive propagation and reintroduction into natural habitats where appropriate.

HABITAT CONSERVATION

Unfortunately, forest conservation is still a low economic and/or political priority, because it does not translate easily into immediate benefits. At present, only a fraction of the money provided each year by institutions such as the World Bank and the United States Agency for International Development (USAID) goes to tropical forest conservation. Replanting and management of valuable forest trees could make sustained utilization of forests possible, but in practice such activities are uncommon and small scale (U.S. Department of State, 1980; Office of Technological Assessment,

1984). More effort needs to be expended to properly manage parks and reserves and to create new ones where necessary. Finding ways of making parks economically self-sustaining and perhaps profitable are also important, since this will encourage local and national governments to support more actively park creation and maintenance.

The Atlantic Forests

The area of highest priority in the Neotropics is obviously the Atlantic forest region of eastern Brazil, with two endemic genera and nine endemic species of endangered primates. This region is the agricultural and industrial heart of the country, as well as its population center. Consequently, only about 1 to 5% of its original forest cover (estimated originally to be over 1,000,000 square km) now remains.

Since 1979, the World Wildlife Fund (WWF) has supported a broad program devoted to the Atlantic forest region, with numerous projects addressing various aspects of conservation. The whole WWF program grew out of a small primate survey initiated in 1979 and primate conservation continues to be a major component of this Atlantic forest program. The bulk of this work is being carried out by a team of some 20 Brazilian students and researchers, with some North American and British participation (Mittermeier 1987a).

Amazonia

The area of second highest priority is Amazonia. By far the largest tropical forest region in the world and home to the majority of New World monkey species (Mittermeier and Cheney 1987), there are as many as 13 sympatric species in some areas.

At the generic level, the primates of greatest concern in Amazonia are *Ateles* and *Lagothrix*. Neither of these are in nearly as critical a situation as are *Brachyteles* and *Leontopithecus*, which are facing the danger of becoming extinct as genera, but both spider and woolly monkeys are disappearing rapidly, and many are already gone from much of Amazonia. The next most endangered Amazonian monkeys are several subspecies with very restricted ranges in areas of heavy development. Included in this latter group are the two southern subspecies of the bearded saki (*Chiropotes satanas satanas* and C. s. *utahickii*), the white uakari (*Cacajao calvus calvus*), the pied bare-face tamarin (*Saguinus bicolor bicolor*) and the white marmoset (*Callithrix argentata leucippe*). All four of these animals are endemic to Brazilian Amazonia (Rylands and Mittermeier 1982).

Though to a lesser degree than the Atlantic forests and Amazonia, the forests of Central America, the cloud forests of northern Peru and Colombia and the drier forests of Argentina, Paraguay and Bolivia must also be considered critical areas for primate conservation. The strategy for these regions should focus on protecting highly endangered species (e.g., the yellow-tailed woolly monkey and the cotton-top tamarin), and on establishing and maintaining appropriate protected areas wherever possible.

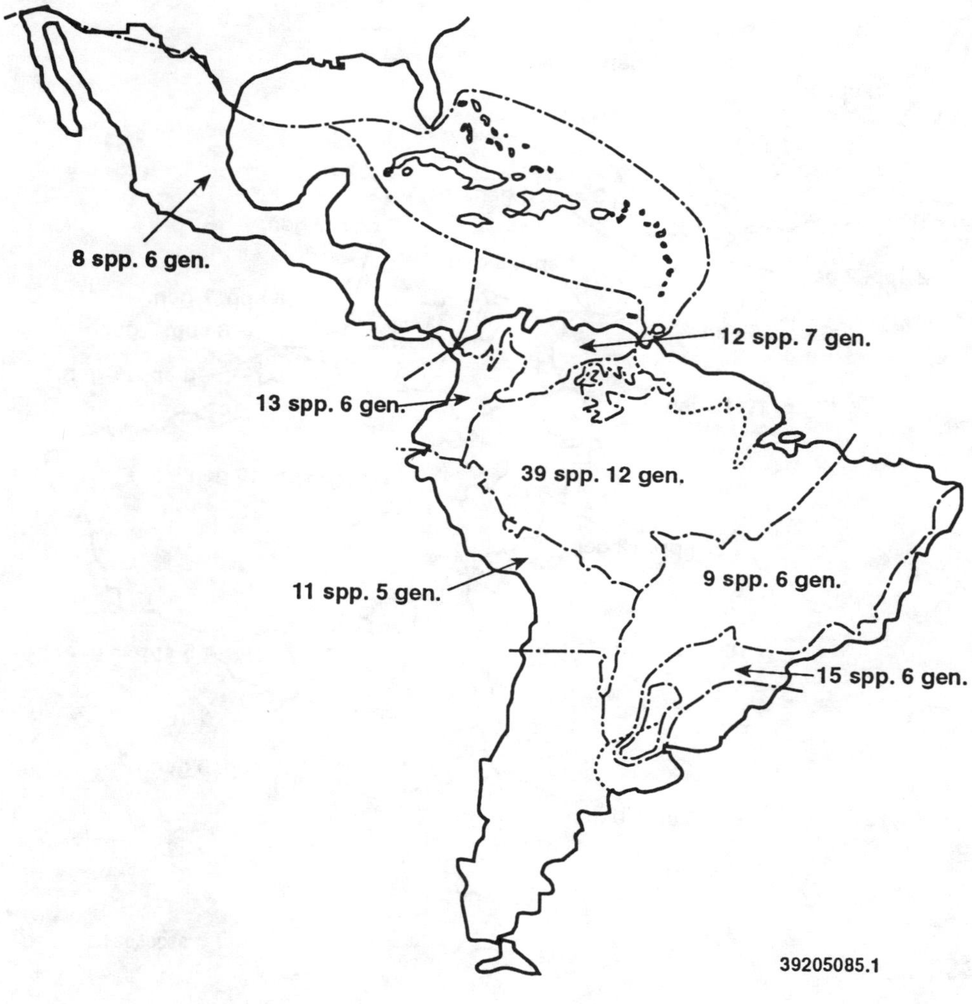

FIGURE N° 1. Number of Neotropical Primate Species and Genera in Each Major Phytogeographic Region

FIGURE N° 2. Neotropical Primate Species and Genera by Country

PROTECTION OF THE PRIMATES THEMSELVES

Although protection of habitats is the major target of Neotropical primate conservation, it is also necessary to ensure that the animals themselves are protected. This means ending all illegal and otherwise destructive traffic in primates, both in the country of origin and in the importing countries.

Commerce in wild primates can only be curtailed if the consumer countries refuse to trade in them. The economic incentives for capturing primates are high and will remain so as long as individuals, zoos and medical institutions are willing to take part in the live primate trade. The decline of the pet trade in Neotropical primates during the last decade can be attributed mainly to the imposition of export bans by source countries, enforcement of import restrictions by user countries and a decreased demand for primates in biomedical research (Mack and Mittermeier 1984).

The international trade in primates has been curtailed significantly since the drafting of the Convention on International Trade in Endangered Species of Wild Flora and Fauna (CITES) in 1973 and the imposition of a number of export bans by source countries. Parties to CITES agreed to ban commercial trade in endangered species and to monitor the trade of others that may become endangered. To date, all the Latin American countries except Mexico have ratified or acceded to CITES. Although the convention provides valuable guidelines for the control of wildlife trade, it s only as effective as the countries that implement it, and some countries follow its requirements more carefully than others. Loopholes that permit, for example, the legal sale of captive-born animals or the importation of animals captured in specific countries, contribute to the traditional forms of illegal trade, bribery, corruption and the intentional misidentification of species (Wolfheim 1983).

Although the CITES convention is occasionally violated, it has effected a marked reduction in the number of non-human primates imported to developed countries each year. For example, in 1968, before the ratifying of the CITES accord, the United States imported 113,714 monkeys and apes; by 1983, this number had dropped to 13,148 (Wolfheim 1983; Mack and Mittermeier 1984). Efforts have also been made to set guidelines for the use of endangered and wild-caught primates in biomedical research. In 1981, the IUCN Primate Specialist Group formulated a policy on the use of primates in biomedical research which recommended that endangered, vulnerable and rare species be used for medical research only when they are obtained from existing, self-sustaining captive breeding colonies. It is important that research institutions that use primates become aware of the conservation status of the species they use, that they use primates as prudently as possible, and that they make every effort to breed the primates that they need.

EDUCATION AND PUBLIC AWARENESS

For any primate conservation effort to succeed, there must be public awareness of the need for conservation and of the importance of primates as both a national heritage and a resource, especially in the countries where they occur. If tropical forest habitats are to be saved, efforts will have to be directed as much towards education as to the protection of the animals and their habitats; as much towards creating economic incentives as to developing parks and reserves; as much towards increasing the realization that agriculture, reliable water supplies, and hydroelectric power all ultimately depend on forest conservation.

Several successful education campaigns are taking place in Brazil, Peru and Colombia. The muriqui is Brazil's largest endemic mammal, and as such, it is most appropriate as a symbol for an Atlantic forest conservation program and for the Brazilian conservation movement as a whole. To publicize the plight of this species and its habitat, the World Wildlife Fund launched a Muriqui Campaign in 1981 in collaboration with the Brazilian Conservation Foundation, the Rio de Janeiro Primate Center and the University of Minas Gerais. This campaign includes lecture programs in the cities and in the interior, development of museum exhibits and a film entitled "Cry of the Muriqui," and distribution of t-shirts, posters, stickers and other educational materials. The campaign has succeeded in getting a great deal of media coverage for the muriqui. The film has already been shown hundreds of times in Brazil, especially in small towns and villages near muriqui habitat. All of this activity has led to a general increase in conservation awareness in Brazil, manifested by the appearance of the muriqui on the cover of a Brazilian phone book, on postage stamps and in frequent magazine and newspaper articles.

In Peru, several projects are underway on behalf of the highly-endangered yellow-tailed woolly monkey [*Lagothrix flavicauda*]. An education and public awareness campaign is now being carried out by a Peruvian researcher and it involves the same kind of materials used in Brazil. In addition to the education program, two potential sites for new protected areas for *I. flavicauda* are also being investigated.

FIELD RESEARCH

Scientific research makes important contributions to conservation also, such as establishing where and at what densities primates occur, and in determining the size of reserves required to maintain viable breeding populations. Behavioral studies are important in establishing the reproductive and social parameters that are necessary for successful captive breeding. The ability to establish propagation programs for species such as the muriqui will ultimately derive from field studies of their behavior. Behavioral and ecological field research also provides a scientific presence that in itself offers a degree of protection and prestige to a given area. In some cases, a scientific research presence has provided an incentive for the establishment of national parks or reserves where none existed previously, such as the privately-owned Fazenda Monte Claros in Minas Gerais, Brazil, which supports one of the most important populations of *Brachyteles*.

CAPTIVE BREEDING AND REINTRODUCTION INTO THE WILD

For certain endangered species, captive breeding programs can be an important component of an overall plan to save an endangered primate. Although breeding programs offer the most important hope for some species, they should not replace, or even be regarded as equivalent to, efforts to conserve wild primate populations.

The captive breeding program for *Leontopithecus rosalia* has been highly successful, and the results of reintroducing this animal into the wild are encouraging (Kleiman et al. 1986). Similar captive breeding efforts are beginning with the endangered black lion tamarin [*Leontopithecus chrysopygus* (Padua 1986)], the golden-headed lion tamarin (*Leontopithecus chrysomelas*) and several other endangered Atlantic forest species at the Rio de Janeiro Primate Center and elsewhere.

PLANS FOR THE FUTURE

It is important to point out that conservation efforts such as those mentioned above are not being carried out in isolation, but rather as part of a broad international effort that has been coordinated by the IUCN/SSC Primate Specialist Group since 1977. The goal of this group is to maintain the current diversity of the Order Primates, with emphasis on both survival of key endangered species wherever they occur, and protection of areas of particularly high primate diversity and abundance.

To achieve this goal, the Primate Specialist Group is developing regional Action Plans for Primate Conservation for each of the four major regions of the world in which primates occur: Asia, Africa, Madagascar and the Neotropics. The Action Plan for African Primate Conservation (Oates 1986) and for Asian Primate Conservation (Eudey 1987) are completed, and the Madagascar and Neotropical Action Plans will follow soon.

In Brazil, one goal has already been largely achieved; ninety percent of the projects that the World Wildlife Fund supports are being carried out by Brazilians, and it is hoped that a similar situation will result in the rest of the Neotropical region within the next 10 years. Hopefully, all the Neotropical countries can be aided and encouraged to the point where they will carry out their own conservation efforts, to be aware of the extent of the problems and to develop the economic capacity to provide the solutions. If this can be done, then there is no reason why all Neotropical primate species should not continue to survive into the next century, and beyond.

REFERENCES

Boinski, S. 1986. Evaluation of the status of the squirrel monkey (*Saimiri oerstedii*), in Costa Rica. *Prim. Conserv.* **6**, 15-16.

Eisenberg, J. F. 1979. Habitat, economy and society: Some correlations and hypotheses for the neotropical primates, pp. 215-261. In *Primate Ecology and Human Origins*, I. S. Bernstein and E. O. Smith, eds. New York, NY, USA: Garland Press.

Eudey, A. 1987. IUCNISSC Primate Specialist Group Action Plan for Asian Primate Conservation: 1986-1990, 41 pp. Washington, DC, USA: World Wildlife Fund/IUCN/UNEP.

Fearnside, P. M. 1982. Deforestation in the Brazilian Amazon: How fast is it occurring? *Interciencia* **7**, 82-88.

Hernandez-Camacho, J. and Cooper, R. W. 1976. The nonhuman primates of Colombia, pp 35-69. In *Neotropical Primates. Field Studies and Conservation*, R. W. Thorington and P. G. Heltne, eds. Washington, DC, USA: National Academy of Sciences.

International Institute for Environment and Development and World Resources Institute. 1987. *World Resources 1987*. New York, NY, USA: Basic Books, Inc.

Kleiman, D. G., Dietz, I. A., Dietz, J. M., Beck, B. B., and Ballou, J. D. 1986. Reintroduction of golden lion tamarins: A model for the conservation of an endangered primate. *Primate Report* **14**, 159.

Lanley, J. P. 1982. Tropical Forestry Resources. Rome, Italy: FAO Forestry Paper #30.

Mack, D, and Mittermeier, R. A., eds. 1984. The Primate Trade. Washington, DC, USA: World Wildlife Fund-US and TRAFFIC-USA.

Mallinson, J. J. C. 1984. Golden-headed lion tamarin contraband a major conservation problem. *IUCNISSC Primate Specialist Group Newsletter*, **4**, 23-25.

Melillo, J. M., Palm, C. A., Houghton, R. A., and Woodwell, G. M. 1985. A comparison of two recent estimates of disturbance in tropical forests. *Environ. Conserv.* **12**, 37-40.

Mittermeier, R. A. 1977. Distribution, Synecology and Conservation of Suriname Monkeys, Ph.D. dissertation. Cambridge, MA, USA: Harvard University.

Mittermeier, R. A. 1987a. Framework for primate conservation in the neotropical region, pp. 305-320. In *Primate Conservation in the Tropical Rain Forest*, C. W. Marsh and R. A. Mittermeier, eds. New York, NY, USA: Alan R. Liss, Inc.

Mittermeier, R. A. 1987b. Effects of Hunting on Rain Forest Primates, pp. 109-146. In *Primate Conservation in the Tropical Rain Forest*, C. W. March and R. A. Mittermeier, eds. New York, NY, USA: Alan R. Liss, Inc.

Mittermeier, R. A. 1987c. Monkey in peril. *Nat. Geogr. Mag.* **171**, 387-395.

Mittermeier, R. A. 1987d. Primate conservation priorities in the neotropical region, pp. 221-240. In *Primates: The Road to Self-Sustaining Populations*, K. Benirschke, ed. New York, NY, USA: Springer Verlag, Inc.

Mittermeier, R. A. and Coimbra-Filho, A. F. 1977. Primate conservation in Brazilian Amazonia, pp. 117-166. In *Primate Conservation*, Prince Rainier of Monaco and G. H. Bourne, eds. New York, NY, USA: Academic Press.

Mittermeier, R. A., Coimbra-Filho, A. F., Constable, I. D., Rylands, A. B., and Valle, C. 1982. Conservation of primates in the Atlantic Forest Region of Eastern Brazil. *Int. Zoo Yearb.* **22**, 1-17.

Mittermeier, R. A. and Cheney, D. L. 1987. Conservation of primates and their habitats, pp. 477-490. In *Primate Societies*, D. L. Cheney, R. M. Seyfarth, B. Smuts, T. T. Struhsaker, and R. W. Wrangham, eds. Chicago, IL, USA: University of Chicago Press.

Mittermeier, R. A., de Macedo, H. and B. A. Luscombe, 1975. A woolly monkey rediscovered in Peru. *Onyx* **13**, 41-16.

Mittermeier, R. A., de Macedo, H., Luscombe, B. A., and Cassidy, J. 1977. Rediscovery and conservation of the Peruvian yellow-tailed woolly monkey (*Lagothrix flavicauda*), pp. 95-115. In *Primate Conservation*, Prince Rainier of Monaco and G. H. Bourne, eds. New York, NY, USA: Academic Press.

Myers, N. 1976. An expanded approach to the problem of disappearing species. *Science* **193**, 198-202.

Myers, N. 1984. *The Primate Source*. New York, NY, USA: Norton.

Myers, N. 1987. Trends in the destruction of rainforest, pp. 3-22. In *Primate Conservation in the Tropical Rain Forest*, C. W. Marsh and R. A. Mittermeier, eds. New York, NY, USA: Alan R. Liss, Inc.

Neyman, P. F. 1977. Aspects of the ecology and social organization of free-ranging cotton-top tamarins (*Saguinus oedipus*) and the conservation status of the species, pp. 39-71. In: *The Biology and Conservation of the Callitrichidae*, D. J. Kleiman, ed. Washington, DC, USA: Smithsonian Institution Press.

Oates, J. F. 1986. IUCN/SSC: Primates Specialist Group Action Plan for African Primate Conservation: 1986-1990, 41 pp. Washington, DC, USA: World Wildlife Fund.

Padua, C. V. 1986. Captive plus wild breeding, an integrated plan for the conservation of the black lion tamarin (*Leontopithecus chrysopygus*). *Primate Report*, **14**, 158.

Peres, C. 1987. Conservation of Primates in Western Brazilian Amazonia. A Progress Report to the World Wildlife Fund-U.S. Washington, DC, USA: World Wildlife Fund.

Rosenberger, A. I. and Coimbra-Filho, A. F. 1984. Morphology, taxonomic status and affinities of the lion tamarins, *Leontopithecus* (Callitrichinae: Cebidae). *Folia Primatol.* **42**, 149-179.

Rylands, A. B. and Mittermeier, R. A. 1982. Conservation of primates in Brazilian Amazonia. *Int. Zoo Yearb.* **22**, 17-37.

Skorupa, J. 1986. Effects of Selective Timber Harvesting on Rain Forest Primates in Kibale Forest, Uganda, Ph.D. dissertation, Davis CA, USA: University of California.

Soini, P. 1982. Primate conservation in Peruvian Amazonia. *Int. Zoo Yearb.* **22**, 37-47.

Tardiff, S. D. and Colley, R. 1987. International Cotton-Top Tamarin Studbook. Oak Ridge, TN, USA: Oak Ridge Associated Universities.

Thornback, J. and Jenkins, M. 1982. IUCN Red Data Book. Gland, Switzerland: International Union for the Conservation of Nature and Natural Resources.

U.S. Department of State. 1980. The World's Tropical Forests: A Policy Strategy and Program for the United States, publication #9117. Washington, DC, USA: Department of State,

U.S. Office of Technological Assessment. 1984. Technologies to Sustain Tropical Forest Resources, Washington, DC, USA: Congress of the United States.

Wolfheim, J. G. 1983. *Primates of the World*. Seattle, WA, USA: University of Washington Press.

Yost, J. A. and Kelley, P. M. 1983. Shotguns, blowguns, and spears: The analysis of technological efficiency, pp. 189-224. In *Adaptive Responses of Native Amazonians*, R. B. Hames and W. T. Vickers, eds. New York, NY, USA: Academic Press.

THE STATUS OF THE NATURAL POPULATIONS OF PRIMATES AND THE BENEFITS OF SUSTAINED CROPPINGS

L. Moya
F. Encarnación
R. Aquino
J. Tapia
C. Ique
P. Puertas

Center for Reproduction and Conservation of Non-human Primates of the Peruvian Primatological Project
Veterinary Institute of Tropical and High Altitude Research
IVITA
Universidad Nacional Mayor de San Marcos
P.O. Box 621
Iquitos, Peru

I. INTRODUCTION

The importance of the nonhuman primates has been amply demonstrated in the scientific world, particularly in the fields of biomedicine and ecology. In biomedicine their importance has been seen through their use as experimental models for the benefit of human health, physiology, biology, pharmacology, and psychiatry, among others. In ecology they fulfill an important role in the propagation and dispersion of seeds, participating in the maintenance of the equilibrium and complexity of the tropical ecosystem where they live.

Prior to 1973 Peru was one of the principal exporters of primates with no criteria for management. Official data report exportation of 339,080 live individuals between 1964 and 1974. Of the total, 77.8% were of the genus *Saimiri*. For the same period there were 15 times more monkeys sacrificed as a consequence of illegal exports, mortality in the period prior to export, and hunting for food in the rural environment.

At the end of 1973, the Peruvian Government declared a closed season on Amazonian wild fauna for an indefinite time, at the time of the initiation of the activities of the Peruvian Primatological Project "Manuel Moro Sommo."

At present the contributions are important for the understanding of the biology and ecology of mainly those species that are of biomedical interest and also of their sustained management in their natural habitat and in captivity. The experiences of this project signify an important contribution in the sustained utilization of the primates envisaged among the principles of the World Strategy for Conservation (International Union for the Conservation of Nature, 1980).

In this document we offer aspects of the current status of the natural populations of the primates with emphasis on those distributed in the Peruvian Northeast. It also includes the benefits of the sustained harvesting of *Saguinus mystax* and *Aotus nancymaei*, employed as models in research on viral hepatitis and malaria, respectively. The experiments have permitted assessment of the species and their habitats, the later considered marginal in traditional agricultural and livestock terms.

II. FORESTS AND PRIMATE POPULATIONS

The forest constitutes an ecosystem characterized by biological diversity, regulated by the edaphic components, the gradients of nutrients, and the concentration of edaphic and atmospheric moisture, which respond to climatic factors such as rain, temperature, solar radiation, and wind. In all the biological diversity, the species of primates deserve special attention because of their habits, patterns of behavior, social organization, use of territory, and their distribution in the biogeographical regions, as well as their adaptation and evolution and morphological, anatomical, and physiological affinity to man.

In the forests of the neotropics, primate species are entirely tree-dwelling; they occupy and live among the horizontal and vertical surfaces formed by the continuity or succession of the trees or succession or intermixing of their branches with the lianas, epiphytes, and semiepiphytes, which in rhythms of periodicity or phenology offer them the principal components of their diet and refuge for rest from the rigors of the climate. In that way they make possible the reproduction and perpetuity of the species of both primates and the plant components. In this delimitation of the manifestations of life, we give notice of the close relationship and dependency between plants and primates for the life of the forest.

From the point of view of the seasonal concentration of water in the soil, in the neotropics we can distinguish two large forest ecosystems: a) highland forests and b) lowland forests. Similarly, there are two other systems determined by the influence of atmospheric moisture distinguished in tropical rain forest and tropical dry forest.

In Peru the natural populations of nonhuman primates are found mainly in the eastern region known as the Selva, which includes 60% of the national territory, almost 75 million hectares. This vast region at altitudes below 400 m above slm, the "selva baja" is inhabited by 28 species and only *Lagothrix flavicauda*, in possible coexistence with *Aotus miconax* and *Callicebus oenanthe* (Hershkovitz, 1988), inhabits the cloud forests at altitudes between 800 and 1100 m above mean slm level in the east-central part. One species, *Alouatta pallita*, inhabits the western forest on the Pacific to the north of the Department of Tumbes, which shares territory with Ecuador.

The croppings in Peru since 1976 occurred mainly in the low land forest adjacent to the large tributaries of the Amazon River in the northeastern part, as well as in the upper basin of the Acre and Tahuamanú Rivers in the southeastern part. The forests can be divided into two large groups according to the preferred habitat of the species still present in Peru, that is, seasonal flooded forest and non-flooded forests. The geographical relief and the plant composition are more important characteristics.

1. Colina forests (non flooded forests)

This habitat is characterized by the hilly physiography, dissected by ravines, with varied slopes that facilitate runoff and drainage of rainwater, generating rivers with white water in the rainy season and clear water during low water. With respect to the physiography, there are terraces and high or low plains, where we differentiate formations and plant communities, such as the palm groves of *Mauritia* and *Phytelephas*, "varillales," "chamisales," and towering trees and forests in galleries. Species such as *Callicebus torquatus* inhabit only the "varillales," whose botanical components include heliophytes and sclerophylls as a result of their adaptation to the white sandy soils. These "varillales" are between the Nanay and Itaya Rivers, currently threatened by agriculture from Iquitos. In addition, *Callicebus torquatus*, associated with *Pithecia*, inhabits the "varillal" forests on the left bank of the central part of the Napo River.

The forests and palm groves among the hills of the Manití, Tamshiyacu, Tahuayo, and Tapiche Rivers contain *Cacajao calvus*, *Ateles*, *Cebus*, *Saguinus*, and others, while the crowns of the trees in galleries are inhabited by *Callicebus*, *Aotus*, *Saguinus fuscicollis*, *S. mystax*, and *Cebuella pygmaea*. The reports on *S. mystax* by Soini and Soini (1982) and Glander (1984) for two different points on the Manití River indicate densities of 25 to 32 individuals/km^2 and 9.7 to 29.5 ind./km^2, respectively, while Ramirez (1984) reports 23.7 to 24.7 individuals/km^2 for the Yarapa River. In southeastern Peru, on the left bank of the Tahuamanú River, (subhumid tropical forest) between 26 and 36 individuals/km^2 of *Saguinus labiatus* have been recorded sharing their territories with S. imperator S. Fuscicollis weddelli, Callimico goeldii, Cebuella pygmaea and others.

2. "Bajial" forests (Seasonal flooded forests)

These formations occupy continuous lands adjacent to the masses of lentic and lotic water, either white or black (Sioli, 1968), so that there has been adaptation to the influence of seasonal variations in the level of water, which on occasion is higher than 4 meters. An appraisal of the sub-basins with forests exposed to flows of black and white water is contained in Encarnación (1985) for the Amazon lowland. Our experiences in forests of the Nanay, Napo, and Tahuayo Rivers (Aquino and Encarnación, 1988) give us densities of 24.2 individuals/km^2 and 33.0 individuals/km^2 for *Aotus nancymae* and *A. vociferans*, respectively. The greatest densities for *Saimiri boliviensis* and *S. sciureus* are also given for the seasonal flooded forests. The presence of 13 species of primates in the Pacaya-Samiria National Reserve (Soini, 1986) which encompasses a greater extension of flood-prone forest shows us the importance of this type of forest as a habitat for primates.

3. The forests: Forestal and livestock approaches

The presence of the forest ecosystem is threatened in the world. The approaches for their utilization constitute colonization or establishment of new human settlements, as an escape to avoid the social problems of the cities and the lack of agricultural land in specific regions of the country.

In Peru as in Bolivia, Brazil, Colombia, and Ecuador, which encompass the Amazon basin, the population density is low--reason to justify the governments to promote human settlements, whose goals are oriented to the extraction of plant products (fruit trees and ornamental and medicinal plants), selective felling of timber-yielding trees, construction of roads and hydroelectric dams, deforestation to plant monocultures of rice, coffee, coca, corn, and oil palms, and for livestock raising, and other activities.

The methods for the use of the land for various crops, including pastures, forest management, and protection (Table N° 1), require the application of adequate ecological techniques and principles, without neglecting to search for benefits to better the living conditions of the human population.

TABLE N° 1. Capacity for better land use by region in Peru (Ha, percent).

	CLEAN CULTIVATION	PERMANENT CROP	PASTURE	FOREST	PROTECTION	TOTAL
COAST	1140000 8,36	4060000 3,64	1622000 11,90	1172000 1,26	10207000 74,84	13637000 100,0
SIERRA	1341000 3,42	20000 0,05	10576000 26,98	2092000 5,34	25169000 64,21	39198000 100,0
SELVA	2421000 3,21	219100 2,89	5718000 7,55	46432000 61,35	18924000 25,0	75686000 100,0
						128521000 100,0

Table N° 1 shows that in the Peruvian selva only approximately 19 million Ha would be designated as protected forests. Among the characteristics that unite these forests is their location primarily on the western side of the Amazon region known as the "cejatle" selva, where the specific diversity of primates is less and their density is low. Similarly, the 46 million Ha suitable for forests are located in the selva bja and the "ceja de" selva, whose physiography does not favor agriculture. However, the selective extraction of wood at irregular times and periods and the lack of assistance and technical orientation in agriculture for the colonists, among other factors, are progressively and gradually eliminating the habitats as well as the individuals of species of large primates at the same rate as for subsistence hunting. The status of the selva assigned to clean cropping, which includes pastures, is a factor in the acceleration of the elimination of primate habitats and individuals through the demand for meat in the human settlements.

These examples exist in the basins of the Alto Mayo and the Central Huallaga and are accentuated in the eastern part of the basin of the Madre de Dios River. The land allotted to use for permanent crops covers only some 2 million Ha; fruit and other trees, such as plantains, cacao, coffee, and citrus, are located in ceja de selva mainly, while the oil palm is promoted in the selva baja, the basin of the Manití River. Of these crops, the fruits edible by man directly on maturing favor and/or enrich the habitats of the primates. But if we consider the greater diversity and population of the primates in the selva baja, it becomes necessary to adopt and promote combinations of agriculture and forestry and of agriculture, silviculture, and grazing in this Amazon plain, in order to achieve adequate use of the land, raising or improving its productivity in order to ensure the life of the components of the wild fauna and biological diversity.

4. Marginal land

Of the 75 million Ha of land in the Selva there are almost 3.5 million Ha (Perú Forestal, 1987) of swampy land and smaller areas of white to gray sand, located mainly in the low land Amazon, among the seasonal flooded forest and non flooded forest. The capacity of these lands for agriculture and livestock raising is minimal, with high investment in the sandy ground. Because of the tree cover, they are considered to be among the lands good for forests; however, reforestation in the swamps still is only an expression for lack of adequate techniques and methods. Thus sustained croppings, in terms of management of species such as *Cebus*, *Saimiri*, *Alouatta*, and *Lagothrix* and other species of wild fauna, are the proper alternative that justifies conservation in terms of profitability. In the comparisons done by Vásquez (manuscript, 1990) in a hectare of "varillal" forest near Iquitos, the harvest of edible fruits, medicinal plants and bark, ornamental plants, and plants for basketry and other purposes without disturbing the nature of the forest was twice as profitable as a hectare of recently cleared land with crops of plantain or manioc.

This information leads us to reflect on the value and importance of sustained harvesting of animals and plants in a marginal forest as opposed to the expectations from the establishment of extensive areas with single crops or from selective extraction of trees valuable for their wood.

5. Status of the primate populations

Since the establishment of the project in November 1972 and after 18 years of activities the advances in the knowledge of the species of nonhuman primates in Peru can be summarized as follows:

As of 1975 only 19 species had been reported (Castro, 1978), many little known by the national scientific community, while currently we can recognize 32 species corresponding to 12 genera. The knowledge of some subespecies, following the studies of O. Hill (1957) and P. Herskovitz (1984, 1987a,b, 1988) among others, is still referential; greater effort is required in explorations to determine the geographical distribution and the critical classes in the taxonomy of the species.

Noteworthy findings include the presence of *Saguinus labiatus* in the southeastern part of the country (Encarnación and Castro, 1978) between the middle ranges of the Acre and Tahuamanú Rivers on the border with Bolivia and Brazil; *Saguinus fuscicollis fuscicollis* on the Tapiche River (Soini, 1979), right tributary of the Amazon River, previously reported only in the basin of the Yurua and Yavarí Rivers in Brazilian territory; and the taxonomic revisions with respect to *Aotus*, *Saimiri*, *Cacajao*, *Pithecia*, and *Callicebus* by Hershkovitz (1983, 1984, 1987a, b, 1988), which have contributed to the clarification of the taxonomic status of those species. Considering *Aotus*, the presence of *A. trivirgatus* is ruled out and *A. nancymae* (a new species since 1983), *A. vociferans*, *A. nigriceps*, and *A. azarae* are recognized; there is information available and registration of *A. miconax*, which is possibly endemic. With respect to *Saimiri*, there is confirmation of the presence of two species with three subespecies, *S. sciureus macrodon*, *S. peruviensis boliviensis* (new subspecies since 1984), and *S. boliviensis boliviensis*.

Studies in the Pacaya-Samiria National Reserve (Soini, 1982) have verified the absence of *Saguinus mystax* in these seasonal flooded forests; it is restricted to the right bank of the Amazon River.

Table N° 2 gives the current status of the species in their natural environment, established on the basis of registration and location, occupation of the territory and their distribution, group size, levels of allopatry and sympatry, as well as the factors of mass felling of areas, selective felling,

TABLE N° 2. Status of the species/subspecies of nonhuman primates in Peru.

Species/Sub Species	Common Name	Distribution Region	Conservation Status*	Source
FAMILY *CALLITRICHIDAE*				
1. *Cebuella pygmaea*	Pygmy marmoset	Amazon, all	C	Soini, 1983a
2. *Saguinus fuscicollis* - *S. f. fuscicollis* - *S. f. illigeri* - *S. f. lagonotus* - *S. f. leucogenys* - *S. f. nigrifrons* - *S. f. weddelli*	Tamarin	Amazon, all	C	Soini, 1983b
3. *Saguinus tripartitus*	Tamarin	Northern Amazon	I	Hershkovitz, 1977 Thorington, 1988
4. *Saguinus nigricollis* - *S. n. nigricollis* - *S. n. graellsi*	Black tamarin	Northern Amazon	C	Hershkovitz, 1977, 1988
5. *Saguinus mystax* - *S. m. mystax*	Moustached tamarin	Northern Amazon	C	Soini, 1982
6. *Saguinus labiatus* - *S. l. labiatus*	Red-bellied tamarin	Southern Amazon	I	Encarnación and Castro, 1978
7. *Saguinus imperator* - *S. i. subgrisescens* - *S. i. imperator*	Emperor tamarin	Southern Amazon	I	Hershkovitz, 1977, 1979; Encarnación et al., 1981
FAMILY *CALLIMICONIDAE*				
8. *Callimico goeldii*	Goeldi's monkey	Central Amazon	R	Encarnación and Castro 1979; Aquino, 1986
FAMILY *CEBIDAE*				
9. *Aotus nancymai*	Owl monkey	Northern Amazon	C	Hershkovitz, 1983; Aquino & Encarnación, 1988
10. *Aotus miconax*	Owl monkey	Northwestern Amazon	I	Hershkovitz, 1983, 1988

* C = common, I = indeterminate, R = rare, E = endangered, V = vulnerable

TABLE N° 2. (Continued) Status of the species/subspecies of nonhuman primates in Peru.

Species/Sub Species	Common Name	Distribution Region	Conservation Status*	Source
FAMILY *CEBIDAE*				
11.*Aotus vociferans*	Owl monkey	Northern Amazon	C	Hershkovitz, 1983; Aquino & Encarnación 1988
12.*Aotus nigriceps*	Owl monkey	Southern Amazon	I	Hershkovitz, 1983; Wright, 1978, 1985
13.*Aotus azarae*	Owl monkey	Southern Amazon	I	Hershkovitz, 1983
14.*Saimiri boliviensis* - *S. b. peruviensis* - *S. b. boliviensis*	Squirrel monkey (Roman arc)	Northern, Central, and Southern Amazon	C	Hershkovitz, 1984
15.*Saimiri sciureus* - *S. s. macrodon* - *C. c. discolor*	Squirrel monkey (Gothic arc)	Northern Amazon	C	Hershkovitz, 1984
16.*Callicebus cupreus* - *C. c. discolor*	Red titi	Northern Amazon	C	Hershkovitz, 1988
17.*Callicebus caligatus*	Red titi	Central Amazon	I	Hershkovitz, 1988
18.*Callicebus brunneus*	Red titi	Southern Amazon	C	Hershkovitz, 1988
19.*Cellicebus oenanthe*	Red titi	Central & Western Amazon	I	Hershkovitz, 1988
20.*Callicebus torquatus* - *C. t. torquatus*	Black titi	Northern Amazon	I	Hershkovitz, 1988; Kinzey, 1982
21.*Pithecia aequatorialis*	Saki	Northern Amazon	I	Hershkovitz, 1987
22.*Pithecia monachus*	Saki	Central & Southern Amazon	I	Hershkovitz, 1987
23.*Pithecia irrorata*	Saki	Southern Amazon	I	Hershkovitz, 1987
24.*Cacajao calvus* - *C. c. ucayallí*	Uakari	Northern Amazon	E	Hershkovitz, 1987

* C = common, I = indeterminate, R = rare, E = endangered, V = vulnerable

TABLE N° 2. (Continued) Status of the species/subspecies of nonhuman primates in Peru.

Species/Sub Species	Common Name	Distribution Region	Conservation Status*	Source
25.*Cebus apella* - *C. a. maranonis* - *C. a. peruanus* - *C. a. macrocephalus* - *C. a. pallidus*	Black capuchin	Amazon, all	V	Hershkovitz, (in press)
26.*Cebus albifrons* - *C. a. cuscinus* - *C. a. unicolor* - *C. a. yuracus*	White capuchin	Amazon, all	V	Hershkovitz, 1949; Terborght, 1981
27.*Alouatta seniculus* - *A. s. seniculus*	Howler monkey	Amazon, all	C	Soini, 1982
28.*Alouatta palliata*	Howler monkey	Dpt. of Tumbes (from Tumbes)	E	Pulido
29.*Ateles belzebuth* *A. b. belzebuth*	Long-haired Spider monkey	Northern & Central Amazon	E	Kellog and Goldman, 1944; Hershkovitz, 1972
30.*Ateles paniscus* *A. p. chamek*	Black spider monkey	Amazon, all	E	Kellog and Goldman, 1944; White, 1986
31.*Lagothrix flavicauda*	Yellow-tailed wooly monkey	Central & Western Amazon	E	Graves and O'Neill, 1980 Leo, 1980; Mittermeier et al., 1975
32.*Lagothrix lagotricha* - *L. l. cana* - *L. l. lagotricha* - *L. l. poeppiggi*	Wooly monkey	Amazon, all	E	Fooden, 1963; Freese, 1975; Durham, 1972

* C = common, I = indeterminate, R = rare, E = endangered, V = vulnerable

establishment of colonies, and others that affect hunting for a source of food. Following those criteria and according to legislation in effect, 6 species are endangered or in the process of extinction, 3 are vulnerable, 1 is rare, 10 are indeterminate, and 12 are common or not in danger.

Some genera with species whose individuals weigh less than one kilogram are widely distributed geographically and have high relative densities in their habitats, and consequently are considered unendangered. The genera with species whose individuals weigh more than one kilogram, such as *Alouatta*, *Ateles*, *Cacajao*, *Cebus*, and *Lagothrix* are found to be vulnerable or endangered, mainly from the pressure of hunting.

III. AREAS OF STUDY AND METHODS

Since 1973, in order to find the species in their habitats and to estimate their densities, the application of censuses in different basins of the Peruvian Amazon region was tried. Since 1976, the censuses were intensified for six species of primates; ecological studies were initiated in order to develop techniques of management of the genera *Saguinus* and *Cebuella* in the basins of the Yanayacu, Tamshiyacu, Tahuayo, Yarapa, Carahuayte, and Tapiche Rivers and elsewhere. Moreover, the liberation and the experimental management of the populations of *Saguinus mystax* on Padre Isla were initiated.

Population censuses of *Aotus nancymae* were conducted periodically from 1981 to 1987 in forests on the Nanay, Napo, and Tahuayo Rivers.

Studies on the sustained cropping of *Saguinus mystax* were carried out in an area of 4 km^2 on the Manití River and in 1.9 km^2 on the Yarapa River, both right tributaries of the Amazon River, and in 8.3 km^2 on Padre Isla by the Amazon River, across from the city of Iquitos. For *Aotus nancymae* the area was 2 km^2 located on the Tahuayo River, right tributary of the Amazon River.

The areas of study include seasonal flooded forests, those on river banks, and upland forests (dry land) (Encarnación, 1985). In all of them primary forests predominate; they are interrupted by areas of migratory agriculture, secondary forests known as "purmas," and areas affected by the removal of wood. On Padre Isla the forest is totally marshy, flooded by the Amazon River in the raining season (March to June). In the areas under study and the zones of influence there are human settlements of mestizo origen where the economic activities are agriculture, fishing, hunting, extraction of wood, and collection of wild fruits.

For the population censuses the method of transects was used (Freese et al., 1982). Travel was on foot in the interior of the forest and in river boats (Canoes) when the censuses covered the borders of the rivers.

For the studies of sustained harvests of *Saguinus mystax* we apply the methods of census by transect (observer-animal), of "chapaneo" and capture of Callitrichidae (Encarnación et al., 1978), and of location of groups in predefined areas (method of maps). For *Aotus nancymae* the method of transects at night and that of capture by isolation of groups were applied.

Saguinus mystax is distributed in areas on the right bank of the Ucayali and Amazon Rivers and the left bank of the Yavarí River; the southern limit is recorded as the place known as "Lagarto" on the right bank of the upper Ucayali River at 10°4' south and 73°48' east (Hershkovitz, 1977).

Aotus nancymae occurs on the right banks of the Marañón and Amazon Rivers and both banks of the Ucayali River up to approximately 07°00' south; it also reaches to the north of the Marañón River in the enclaves between the lower and middle parts of the Tigre and Pastaza Rivers.

IV. BENEFITS OF SUSTAINED CROPPING

Activities have been carried out since 1976, through the coordinated actions of the official sectors of the Government of Peru and of the Greater National University of San Marcos, with the support and the technical cooperation of the Pan American Health Organization/World Health Organization.

The harvesting of six species is being carried out, namely: *Aotus nancymaei*, *A. vociferans*, *Saguinus labiatus*, *S. mystax*, *Saimiri boliviensis*, and *S. sciureus*. An example of the results obtained with *Aotus nancymai* and *Saguinus mystax* is presented below.

1. Sustained Cropping

Saguinus mystax

This is the first species in the management of primates in natural areas in Perú. Evaluations of the effects of the capture and population recovery took place in the forests adjacent to the Manití and Yarapa Rivers.

Glander et al. (1984) reported that in the first area the cropping had a minimal effect on the populations, whose recovery to preharvest levels, that is, to 5.0 groups/km^2 or 29.5 individuals/km^2, occurred in the five years after the initial cropping (Table N° 3).

TABLE N° 3 Comparison of the population densities of *Saguinus mystax* and *S. fuscicollis* in the study area, Fundo Los Angeles - Río Manití. Period, 1976-1982.

YEAR	S. mystax		S. fuscicollis		REFERENCE
	Grp./Km2	Ind./Km2	Grp./km2	Ind./km2	
1976 - 77	5.0	25.0	4.7	28.0	Soini 1981(a)
1982	5.3	31.3	3.4	24.1	Glander et al. 1982

In forests on the Yarapa River similar results were obtained. The precapture density was 6.3 groups/km^2 or 32.1 individuals/km^2; 66% were harvested in 1978. The remaining individuals, 21 in number, were regrouped into three groups, so that in 1981, the population showed an active recovery of 124%, compared to the population remaining in 1978, and reached a density of 24.7 individuals/km^2. In 1987, six years afterwards, the population density in the area exceeded the initial population of 1978 by 25%, thus indicating complete recovery. That same year a second harvest of 63% was carried out. In 1989 the evaluation reported the progressive recovery of 25% (Table N° 4).

TABLE N° 4. Census data on the population recovery of *Saguinus mystax* and *S. fuscicollis* in the Yarapa River study area from 1978 to 1989.

	S. mystax		*S. fuscicollis*		
YEAR	TOTAL POPUL.	POP. CHANGE %	TOTAL POPUL.	POP. CHANGE %	REFERENCE
1978(1)	61	-	63	-	Moya et al.,1979
1978(2)	21	-66	57	-9,5	
1981	47	124	50	-12	Ramírez, 1984
1982	45	- 4	59	18	
1987(1)	76	69	73	24	Moya and Puertas, 1988
1987(2)	28	-63	73	24	
1989	35	25	46	-37	Puertas, 1990

(1) Before capture. (2) After capture.

Analysis of the population structure of 10 complete groups captured in 1987 showed a composition of 64.7% adults, 7.5% subadults, 22.6% juveniles, and 5.7% infants. This is very similar to the structure of natural populations registered for the Manití, Tapiche, and Tahuayo Rivers (Soini, 1982). Moreover, the proportion of immature individuals to adults was 1.0:0.56. This relationship indicates a trend toward stabilization as a mature population; the ratio reported for the natural populations on the Manití and Tapiche Rivers is 1.0:0.5 and that for the Tahuayo River is 1.0:0.3 (Soini, 1982). These data suggest a vigorous population, with successive reproductive events in 8 and 9 months. The male/female ratio for adults was slightly favorable for males at 1.0:0.89.

This result is similar to that for 1981 (Ramírez, 1984), but it differs from that of 1978 with a greater proportion of females (16%). New results obtained in 1981 and in 1987 are similar to those reported for the Tapiche, Manití, and Tahuayo Rivers (Soini, 1982), which would explain the demographic phenomenon in adult populations affected by harvesting in which the recovery of the females is progressive to slightly superior to the males; that is considered characteristic of a stable natural population.

Ramírez (1984) estimated the maximum population increase of *S. mystax* at the third year after the first harvest. This model requires an adjustment, since the carrying capacity of the area under study in 1987 reached 76 individuals, while in 1978 it was 61 individuals. That could mean that the maximum rate of increase was lower. Therefore, a longer time will be required for the maximum harvest; however, we estimate that the maximum harvests should be carried out four or five years after the initial harvest.

The experiment of the introduction of *S. mystax* on Padre Island was originated as a breeding alternative under natural conditions but limited by a physical barrier, the Amazon River. The initial population released prior to 1980 consisted of 75 individuals. At the beginning, adaptation and population growth were slow; subsequently, this growth was greater, yielding a total population of 161 animals, which means an average growth of 12.8% in 9 years (Table N° 5).

TABLE N° 5 Initial population and annual growth of the population of *Saguinus mystax* on Padre Isla, Loreto, Peru, from 1980 to 1989.

				COMPOSITION			
YEAR	NO. OF GROUPS	GROUP SIZE	NO. OF INDIVIDS.	ADULTS-SEMIADS.-JUVS.	IN-FANTS	TOTAL INCREASE (%)	INFANTS %
1980	18	4,2	75	75	-	-	-
1981	14	5,9	82	77	05	9,3	6,1
1982	17	4,9	84	82	02	2,4	2,4
1983	16	5,7	91	77	14	8,3	15,4
1984	-	-	-	-	-	-	-
1985	15	6,5	98	98	-	-	-
1986	15	8,2	123	109	14	25,0	11,4
1987	15	8,4	126	114	12	2,4	9,5
1988(a)	17	9,0	153	134	19	21,0	12,4
1988(b)	17	7,6	129	122	07	-15,7	
1989	19	8,5	161	138	23	24,8	14,3

In 1988 the selective harvest of 32 individuals was carried out. This first harvest was successful and has impacted positively on the reproduction of the species (Table N° 5). Actions for the protection of the forests on the island and enrichment with plants of dietary importance for *S. mystax* allow the reproductive levels of the species to increase.

For the management of the natural populations of *Saguinus mystax* the following sequences have been considered:

a) The execution of precapture censuses in different river basins. These reported an average population density of 30 individuals/km^2 or 6 groups/km^2 (Table N° 6).

b) The development of a technique for capture using the "Saguinus" trap for selective manual use.

c) The number of animals to harvest was determined on the basis of the population density, with 50% to 70% of the population or 3 to 4 groups affected and a remainder of 30% to 50% of the population or 2 to 3 groups, in order to facilitate population recovery. Whenever possible complete groups were captured in order to avoid interference with the social organization of the groups.

TABLE N° 6 Population densities of *Saguinus mystax* in Peru.

STUDY AREAS	GRP./KM2	IND./KM2	REFERENCE
Río Manití	4,6	23,0	Soini 1976
Río Tahuayo	6,5	32,0	Soini 1977
Río Ucayali (J.Herrera)	6,0	36,0	Aquino 1978
Río Yanayacu	5,0	25,0	Tapia 1977
Río Carahuaite	5,3	26,5	Tapia 1978
Río Yarapa	6,3	32,1	Moya et al 1978
Río Orosa	5,7	28,5	Tapia 1979
Río Cochiquina	6,9	34,5	Moya 1981
Río Pichana	7,7	38,5	Moya 1981
Río Manití	5,3	31,5	Glander et al 1982
Río Manitk	5,1	29,9	Puertas 1988
	& = 64,4 X = 5,9	337,3 30,7	

Varaition: 23,0 to 38,5 ind./km²

d) Rotation of cropping areas by basin at intervals of 4 to 5 years or more. For example, harvests were carried out on the Manití River in 1976, 1980, and 1988 and on the Yarapa River in 1978 and 1987.

e) Alternation of harvest areas in the same river basin, leaving areas untouched. A harvest area included a stretch of 10 to 15 km along the river and from 1 to 5 km toward the interior of the forest on one or both banks of the river. In a river basin usually from two to three harvest areas were established.

Aotus nancymae

Evaluations of the effects of the harvest were initiated in 1981; the density prior to capture was estimated at 9.5 groups/km^2 or 35.5 individuals/km^2. That same year 59% of the population was trapped; the 29 remaining individuals formed 7 complete or incomplete groups.

In 1984 the repopulation reached 169%, that is, 49 individuals, exceeding the first preharvest density by 3.5 individuals/km^2. That same year 84% of the population was harvested, leaving only 12 individuals. In 1987 the repopulation reached 433%; however, the density was lower by 3.5 individuals/km^2 than in 1981 and 7.0 individuals/km^2 less than in 1984. This suggested that the population had not achieved stabilization. The harvest of almost the entire population does not appear to have had negative effects on the capacity for repopulation.

The mechanisms of repopulation in 1981 have been attributed to the birth rate and the immigration of neighboring groups. The presence of immature individuals in at least five remaining groups is evidence of the birth rate.

In the process of repopulation in the second harvest of 1984, the most important contribution evidently was the immigration of neighboring groups, while the birth rate played a secondary role.

2. Benefits for the forest

One of the strategies for the protection of the populations of primates in Peru consists of the establishment of conservation units or national parks.

At present there are five protected units with populations of primates. Manú National Park, located in the southeastern part of the country, covers 1,532,000 Ha and protects 13 species of primates; Tingo María National Park, covering 18,000 Ha in the west-central part of the Amazon region, protects five species of primates; Abiseo River National Park, located in the west-central part of the Amazon region and with an area of 274,520 Ha, protects mainly *Lagothorix flavicauda*, an endemic species that inhabits the cloud forests in the ceja de selva; Yanachaga Chenillén National Park, located in the central selva and with an area of 122,000 Ha, protects seven species of primates; and the Pacaya-Samiria National Reserve, located in the northern half of the Amazon region in the selva baja and with an area of 2,080,000 Ha, protects 13 species of primates (Table N° 7).

These protected areas cover 3,026,520 Ha, equivalent to 4% of the Amazon region, and protect 17 species. If we consider the total diversity, approximately 50% of the recognized subspecies are protected, with the other 15 species not included (Table N° 8) which justifies the creation of new conservation units.

Currently, the Tamshiyacu-Tahuayo Community Reserve is being created; it is located in the southeastern part of the Department of Loreto and has an area of 322,500 Ha. In this effort protection is sought for 13 species of primates, among them *Saguinus mystax* and *Cacajao calvus*, which are still not represented in the areas protected by the state. Moreover, the creation of the National Park of Loreto on the border with Colombia is being promoted in order to protect species whose distribution reaches to the northeastern part of the country as is establishment of a community reserve in the basin of the Acre and Tahuamanú Rivers, Madre de Dios, to protect 12 species of primates, among them *Saguinus labiatus* and *Callimico goeldii*.

In the system of protected areas, through the application of census techniques and evaluations, trapping, transport, and other means, in accordance with the need to use the species of primates, sustained cropping are applicable--those that will maintain the reason for such areas oriented toward the greatest yield from the forest.

Sustained cropping also guarantees the equilibrium of life in the forest through the ease of recovery of the populations of the affected species, assuming the absence of hunting.

TABLE Nº 7. Nonhuman primate species protected in conservation units in Peru.

Species	Conservation Unit				
	Manu National Park	Tingo Maria National Park	Abiseo River National Park	Yanachaga Chemillen National Park	Pacaya Samiria National Park
Cebuella pygmaea	X	X			X
Saguinus fuscicollis	X	X		X	X
Sagunius imperator	X				
Callimico goeldii	X				X*
Aotus nancymai					X
Aotus nigriceps	X			X	
Saimiri boliviensis	X	X		X	X
Saimiri sciureus					X*
Callicebus brunneus	X				
Pithecia monachus	X				X
Cebus apella	X	X		X	X
Cebus albifrons	X		X		X
Alouatta seniculus	X	X		X	X
Ateles belzbuth			X		X
Altes paniscus	X			X	X
Lagothris flavicuda			X		
Lagothris lagothricha	X			X	X

*Still not confirmed *in situ*.

TABLE N° 8. Nonhuman primate species not protected in the conservation units in Peru.

Saguinus tripartitus
Saguinus nigricollis
Saguinus mystax
Saguinus labiatus
Aotus miconax
Aotus vociferans
Aotus azarae
Callicebus cupreus
Callicebus caligatus
Callicebus oenanthe
Callicebus torquatus
Pithecia aequatoriales
Pithecia irrorata
Cacajao calvus
Alouatta palliata

3. Benefits for humankind

The beginning of the harvest carried out since 1976 in Peru is the beginning of the experiment of sharing the benefits between the rural inhabitant and the urban dweller. At the same time that experiment has promoted and facilitated support in terms of technical cooperation from the developed countries, seeking some patterns of equilibrium and compensation between the countries producing nonhuman primates and the countries using them.

The evidence of this exchange is manifested in the incorporation and integration of the rural inhabitants in the activities such as courses, evaluations, trapping, retention, and transport, among others, with technical fundamentals under the responsibility of specialists in natural resources and management. The construction of a breeding centre and the equipment for its operation, along with the positive results of the management in captivity of the species of interest, also offer opportunities for: a) training students at different levels, b) provision of basic laboratory services for the community, and c) execution of biological and ecological/ethological research.

The actions mentioned signify contributions in environmental education and ecology for the adults in general, which are difficult to carry out at the present time in the hamlets of the Amazon region. On the other hand, they contribute to sociocultural exchange among the countries of the world.

V. CONCLUSION

In summary, one can conclude that in a natural population of *Aotus*, the cropping of a reasonable number of individuals, the cutting down of some trees, and the selective clearing in the capture do not cause negative effects on the population. Moreover, the vegetation shows rapid recovery, even allowing the use of dens as places for "daytime sleep."

The maximum utilization of the potential of the forests of the Amazon region should be carried out with the maintenance, whenever possible, of its natural state, without altering it or destroying it by the removal of one or several of the products, such as it occurs with the selective removal of wood or the construction of agricultural areas. The experiments in the sustained management of the populations of *Saguinus mystax* and *Aotus nancymae* show the great possibilities for carrying out permanent long-term harvests without affecting the forests.

VI. REFERENCES

Aquino, R. 1990. La Fauna Primatológica en áreas de Jenaro Herrera, pp. 122-134. In *La Primatologia en el Perú, Investigaciones Primatológicas (1973 - 1985), Proyecto Peruano de Primatología, "Manuel Moro Sommo"*, N. E. Castro-Rodríguez, ed. Lima, Peru: Dirección General de Forestal y Fauna, Ministerio de Agricultura; Instituto Veterinario de Investigaciones Tropicales y de Altura, Universidad Nacional Mayor de San Marcos; Organización Panamericana de la Salud, Organización Mundial de la Salud.

Aquino, R. and Encarnación, F. 1986. Population and structure of *Aotus nancymai* (Cebidae: Primates) in Peruvian Amazon lowland forest. *Am. J. Primatol.* **11**, 1-7.

Aquino, R. and Encarnación, F. 1988. Population densities and geographic distribution of night monkeys (*Aotus nancymai* and (*Aotus vociferans*) (Cebidae: Primates) in Northeastern Peru. *Am. J. Primatol.* **14**, 375-381.

Aquino, R, Puertas, P. and Encarnación, F. in press. Efectos de las cosechas sobre las poblaciones de *Aotus nancymai* (Cebidae, Primates) en bosque inundable. Iquitos, Peru: Veterinary Institute of Tropical and High Altitude Research, Peruvian Primatological Project.

Castro, N. E. 1978. Diagnóstico de la situación actual de los primates no humanos en el Perú y un plan para su utilización nacional, pp. 20. Lima, Peru: Ministerio de Agricultura y Alimentación

Encarnación, F., Moya, L., Soini, P., Tapia, J., and Aquino, R. 1978. Captura de Callitrichidae *Saguinus* y *Cebuella* en la Amazonía Peruana. *Boletín de Comité Regional de Coordinación del Proyecto Primates* **1**, 1-20.

Encarnación, F. 1985. Introducción a la flora y vegetación de la Amazonía Peruana: Terminología y breve descripción de las formaciones vegetales de la Llanura Amazónica. *Candollea* **40**, 237-252.

Encarnación, F. and Castro, N. 1990. Informe preliminar sobre Censo de Primates no Humanos en el Sur-Oriente Peruano: Iberia e Iñapari (Departamento de Madre de Dios), Mayo 15 - Junio 14, 1978, pp. 57-67. In *La Primatologia en el Perú, Investigaciones Primatológicas (1973 - 1985), Proyecto Peruano de Primatología, "Manuel Moro Sommo"*, N. E. Castro-Rodríguez, ed. Lima, Peru: Dirección General de Forestal y Fauna, Ministerio de Agricultura; Instituto Veterinario de Investigaciones Tropicales y de Altura, Universidad Nacional Mayor de San Marcos; Organización Panamericana de la Salud, Organización Mundial de la Salud.

Freese, C. 1975. A Census of Nonhuman Primates in Peru, pp. 17-71. In: Primate-Censusing Studies in Peru and Colombia, Report to the National Academy of Sciences on the Activities of the Project, (1) AMRO-0719. Washington DC, USA: PAHO

Glander, K., Tapia, J. and Fachín, A. 1984. The impact of trapping and removal of wild populations of *Saguinus mystax* and *Saguinus fuscicollis*. *Am. J. Primatol.* **7**, 89-97.

Hershkovitz, P. 1977. *Living New World Monkeys (Platyrrhini)*, vol. 1, 1117 pages. Chicago, IL, USA: University of Chicago Press.

Hershkovitz, P. 1983. Two new species of night monkeys, genus *Aotus* (Cebidae, Platyrrhini): A preliminary report on *Aotus* taxonomy. *Am. J. Primatol.* **4**, 209-213.

Hershkovitz, P. 1984. Taxonomy of squirrel monkeys, genus *Saimiri* (Cebidae, Platyrrhini): A preliminary report with description of a hitherto unnamed form. *Am. J. Primatol.* **7**, 155-210.

Hershkovitz, P. 1987a. Uacaries, New World monkey of the genus *Cacajao* (Cebidae, Platyrrhini): A preliminary taxonomic review. *Am. J. Primatol* **12**, 1-53.

Hershkovitz, P. 1987b. The taxonomy of South American sakis, genus *Pithecia* (Cebidae, Platyrrhini): A preliminary report and critical review with the description of a new species and new subspecies. *Am. J. Primatol.* **12**, 387-468.

Hershkovitz, P. 1988. Origin, speciation, and distribution of South American titi monkeys, genus *Callicebus* (Family Cebidae, Platyrrhini). *Proc. Acad. Nat. Sci. Phil.* **140(1)**, 240-272.

Hill, W. C. O. 1960. Primates: Comparative Anatomy and Taxonomy IV., *Cebidae*, Part A, a Monograph. Edinburgh, Scotland: University Press, xxii, 523 pp.

Moya, L., Verdi, L., Bocanegra, G. and Rimachi, J. 1979. Análisis poblacionales de *Saguinus mystax* (Spix, 1823) (Callitrichidae) en la cuenca del Río Yarapa, Loreto, Peru. *Boletín de Comité Regional de Coordinación del Proyecto Primates* **2(1)**, 1-44.

Moya, L. and Puertas, P. 1988. Evaluación poblacional y cosecha experimental de *Saguinus mystax* en bosque del Río Yarapa en la Llanura Amazónica Peruana. pp. 1-13. Iquitos, Peru: Proyecto Peruano de Primatología.

Ramírez, M. 1984. Population recovery in the moustached tamarin (*Saguinus mystax*: Management strategias and mechanism of recovery. *Am. J. Primatol.* **7**, 245-259.

Soini, P. 1990. Nota sobré el hallazgo de una sub-especie adicional de *Saguinus fuscicollis* (Callitrichidae, Primates) para el Perú, pp. 314-317. In *La Primatologia en el Perú, Investigaciones Primatológicas (1973 - 1985), Proyecto Peruano de Primatología, "Manuel Moro Sommo"*, N. E. Castro-Rodríguez, ed. Lima, Peru: Dirección General de Forestal y Fauna, Ministerio de Agricultura; Instituto Veterinario de Investigaciones Tropicales y de Altura, Universidad Nacional Mayor de San Marcos; Organización Panamericana de la Salud, Organización Mundial de la Salud.

Soini, P., Soini, M. 1990. Distribución y ecología poblacional de *Saguinus mystax* (Primates, Callitrichidae), pp. 272-313. In *La Primatologia en el Perú, Investigaciones Primatológicas (1973 - 1985), Proyecto Peruano de Primatología, "Manuel Moro Sommo"*, N. E. Castro-Rodríguez, ed. Lima, Peru: Dirección General de Forestal y Fauna, Ministerio de Agricultura; Instituto Veterinario de Investigaciones Tropicales y de Altura, Universidad Nacional Mayor de San Marcos; Organización Panamericana de la Salud, Organización Mundial de la Salud.

Tapia, J., Encarnación, F., Aquino, R., Moya, L. and Soini., P. 1990. Censos poblacionales y sacas periódicas de primates en la Amazonía peruana (1976-1985), pp. 325-341. In *La Primatologia en el Perú, Investigaciones Primatológicas (1973 - 1985), Proyecto Peruano de Primatología, "Manuel Moro Sommo"*, N. E. Castro-Rodríguez, ed. Lima, Peru: Dirección General de Forestal y Fauna, Ministerio de Agricultura; Instituto Veterinario de Investigaciones Tropicales y de Altura, Universidad Nacional Mayor de San Marcos; Organización Panamericana de la Salud, Organización Mundial de la Salud.

Environmental Factors Affecting the Reproduction of Primates in Special Habitats

Thomas L. Wolfle

Institute of Laboratory Animal Resources
National Research Council,
National Academy of Sciences
Washington, D.C.

This presentation focuses on three key elements of effective and humane conservation and management of nonhuman primates: environment, captive research habitats, and individual and species-typical behaviors. A knowledge of each of these is essential. However, in order for them to combine into an effective strategy a fourth element, personnel, must be added to the formula. The management of captive colonies of nonhuman primates requires a knowledge of species-typical behaviors, and the impact that different captive environments have upon enabling or inhibiting these behaviors. The translation of species-typical behaviors, expressed by animals in the wild, to animals in captivity is an obvious starting place but is not without risks. Understanding which behaviors in captivity that are *different* from those in the wild, but not necessariiy abnormal, is an area in which considerable research is needed. Sterotypic behaviors are commonly considered to be indicative of maladaptive responses to the stress of an inadequate captive environment. Many, such as self-mutilation and infanticide, are probable examples. However, other behaviors, such as pacing or sleeping may be behaviors that enable the animal to effectively cope with the new environment. Understanding the dynamics of these responses to captive environments is a challenge requiring collaborative research between behaviorists, primatologists, endocrinologists, reproductive physiologists, technicians, and others. Colony managers charged with the responsibility for humane care and breeding of nonhuman primates must clearly enunciate and document a program that addresses their behavioral and physiological well-being. This program should identify goals and options for achieving the goals, and clearly delegate authority to those who have the professional skills necessary for interpretation and implementation.

Introduction

When Linnaeus gave the name *Primate* to the taxonomic order including man, monkeys, and apes he no doubt was impressed with the diversity of this "first" order of the animal kingdom, which includes behavioral, morphological, and social differences of enormous proportions. Morphologically, nonhuman primates range in size from less than one kilogram to over 100 kilograms; some have prehensile tails; some are adapted for a brachiating, arboreal life-style; others are adapted for ground-dwelling. Behaviorally and socially, they include those who are active at night and those who are active in daylight; those who are solitary and those who are social; those who mate for life and those whose social structure supports harem mating; those who are elaborate nest-makers and those who sleep in improvised settings; and those whose hierarchial structure is determined by matriarchal or patriarchal lines, size and strength, or family status.

These, and other, differences are the basis for the variety of the species-typical behaviors seen in the over forty species of nonhuman primates encountered in captive research environments. They provide enormous challenges to colony managers in establishing proper caging and environments that enable animals to elaborate their species-typical behaviors. It is generally considered that the expression of species-typical behaviors is a desirable manifestation of an animal's ability to adapt effectively within the captive environment. Thus, the cost in energy to an animal to adapt to a new environment is an essential consideration when judging the adequacy of the environment and the amount of stress that it (and the adaptation process) places on the animal. In the final analysis, the animal's ability to adapt to a new environment with minimal ongoing stress is the desired goal, both from the standpoint of animal welfare and for the physiological and biochemical stability essential for research.

It is clear that animals are under some degree of stress whether in the wild or in captivity. For example, stresses associated with prey/predator relationships and with food gathering in the wild are, to some degree, replaced in captivity with stresses associated with boredom, lack of appropriate social interactions, or space restrictions. The goal is not, therefore, to identify and alleviate every source of stress, but to identify behavioral, physiological, and biochemical changes that are indicative of a homeostatic process that is unable to return the animal to a stable baseline, which means that the animal is unable to adapt.

These comments are not intended to be all-encompassing, but to establish a conceptual base upon which to build objectives and actions in accordance with the goals of this first meeting of the Regional Primatology Committee for the Americas. The following discussion amplifies on these introductory thoughts and provides a strategy by which various disciplines may address the issues in detail.

Lumping or Splitting

It sometimes seems that discussions on this topic are inconsistent in that they first stress the importance of identifying the individual variability that exists among animals of a species (splitting), and then discuss species and even families or orders as if there were no important differences (lumping). In fact, both are valid and equally important. There is so much individual variability among, for example, individual squirrel monkeys (*Saimiri sp*) or night monkeys (*Aotus sp*), even within those of the same geographical region or subspecies, that one is constantly reminded of the necessity of learning the individual differences of the behaviors of each animal in a colony in order to judge the responses of *that* animal under a research project or environment. These individual differences are frequently, but not invariably, related to experiences derived from the captive environment and may be manifested as increased or decreased aggression toward cage mates or self, fear and associated distress associated with the proximity of people or restraint for medical or experimental reasons, or increased or decreased appetite or activity. Knowing the baseline level of these or other behaviors prior to placing an animal in a group cage with others, or on a research protocol, may provide early warning when the animal manifests unacceptable levels of stress and indicate that some intervention is needed. It must be emphasized that recognition of when to intervene is often difficult unless these individual differences are known, and even then must be made with professional judgement. Research or animal care technicians are frequently the best individuals to make these decisions, for they are often the most knowledgeable about the animals under their care. Another point must also be made. Alleviation of stress can be addressed from the standpoint of the well-being of the animal or from the reliability of the research data. It may be the case that an animal actively engages in activities (behavioral or physiological) that enable it to cope with the existing environment, but that are incompatible with the research. Intervention for the alleviation of stress may, therefore, be required for reasons of maintaining research stability even when the animal is apparently adapting well to the environment. Depending on the nature of the research, the opposite may also be true - intervention may be necessary for the well-being of the animal even though the research is in no jeopardy. In each of these cases, the goals of the research and judgements of the scientist, veterinarian, and animal care and use committee are essential considerations that should be clearly identified and discussed in an institutional plan.

Having said that about individual differences, species similarities are, nevertheless, worthy of consideration for they form the basic knowledge people have about animals. The differences between macaques and tamarins are understood primarily due to a knowledge of representative species, such as rhesus monkeys (*Macaca mulatta*) and red-bellied tamarins (*Saguinus labiatus*), respectively, irrespective of the individual differences among the members of each species. When one plans on breeding or housing a nonhuman primate species, requirements for housing and maintenance are based primarily on a knowledge of the species, and to a lesser extent on an understanding of the range of individual differences that are likely to found with the group. The desirability of pair-, family-, or large-group housing depends primarily on the species, with the need for alternate housing depending upon the individual differences and conflicts within the group. It is therefore quite obvious that prior to receiving a colony, or a single individual, one should have a detailed understanding of the species.

From the above it should be evident that it is not sufficient to "split" species and define them only as consisting of a group of individual differences. Nor is it sufficient to "lump" all individuals of a species into a single prototype. Clearly, proper humane management and use in research requires a thorough knowledge of both species-typical traits and individual differences.

A Plan of Action - The Institutional Plan

In order to address *what* to do to address the housing, care, and breeding of nonhuman primates, a plan of action for *how* to proceed must be developed. This "institutional plan" has several key components that must be in place before such specifics as environment, habitat, and reproduction can be placed in perspective.

Although any single plan may, over time, become quite inclusive, different institutional plans vary greatly depending on many factors. One issue that is invariant, however, is that a focus on any one parameter, such as reproduction, is not sufficient. In this example, reproduction is the end result of a well managed colony that addresses the well-being of each individual with it. Well-being, not reproduction, is the goal of the plan. While reproduction is an important, even critical, indicator, a plan that only addresses reproductive parameters and fails to consider other environmental, social, nutritional, and husbandry factors is short sighted. Although it may come as a surprise to some, well-being is more than reproduction and freedom from pain and distress. An assessment of the well-being of animals within a captive environment must address many other parameters, most important of which is the motivation and training of those who care for and use them.

The stability of the foundation upon which rests the adequacy of conservation programs, and the management and health of captive nonhuman primates, depends upon the clarity of expression and application of an institutional program that annunciates the goals, responsibility, and lines of authority. Just as it is not acceptable to bring new species of primates into a captive setting without the proper education and training of those who will interact with the animals, neither is it acceptable to initiate or conduct conservation, breeding, or research using primates without a clearly written and communicated policy and goal. This statement, from the senior institutional official involved with the program, should identify a single individual responsible for the entire program who has the authority to commit essential resources of personnel and finances to accomplish the purpose. Because most such programs affect different departments within a university or corporation, the responsible person should occupy a position within the administration who has administrative authority above each of the component departments. The plan should provide a clear statement of the goals of the project with sufficiently detailed bench-marks to permit measurement of the degree of attainment of, or departure from, the goals. The chapter headings in ILAR's *Guide for the Care and Use of Laboratory Animals* are commonly used for plans developed by institutions that use animals in research, testing, or education. Similar guidance for conservation programs may be developed from principles or practices of conservation, such as the International Primatological Society's *International Guidelines for the Acquisition, Care and Breeding of Nonhuman Primates*.

Personnel are the key to conservation, breeding, and management programs. Although facilities are often cited as essential to "accreditable" programs, current wisdom recognizes the value of personnel over that of the bricks and mortar of the facility. This focus on personnel begins with the commitment of the individual responsible for the program and the degree to which he or she communicates this commitment to those who are affected. The various personnel who directly or indirectly affect the animals or their environment are of equal importance. These persons include not only investigators, biologists, technicians, and veterinarians who all have direct interaction with the animals, but also include engineers, secretaries, facility guards, and sanitation personnel who have casual contact or who in one way or another may have cause to enter the animal facility or whose job may be dependent upon the animal facility. The education of this latter group may benefit the animals directly through facility maintenance by building and sanitation engineers, or indirectly through expressions to the public of support for the programs by secretaries and office support personnel. The institutional plan should identify each of these persons and the training or education appropriate to each.

Caging Types and Social Propensities

The first consideration one generally has when planning for the housing of nonhuman primates is whether the animals are to be housed singly, in pairs, or in social groups. Although exceptions do exist, most species of nonhuman primates live in social groups in the wild and will benefit from the opportunity for some degree of social interaction with conspecifics (same species). As stated previously, the bonding force of the social structure may be established through family lines or hierarchically through the brute force of an "alpha" male. They may be tightly formed units or loose aggregates, and may consist of only a pair or a troop of more than a hundred. These species-typical characteristics must be known for each species being managed, bred, or conserved.

Once a housing system is decided upon, a strategy must be developed for introducing animals into it and for monitoring animals within it. It is not sufficient to just "dump" primates into a social group and wait for the babies, although the intentional chaos caused by multiple, simultaneous introductions does seem to reduce aggression by existing colony members over that they show for singly introduced individuals. A method of adaptation of members of the planned social group to one another must be developed in order to minimize aggression and trauma from incompatible animals. Once the social group is formed, frequent observation must ensure that excessively aggressive or traumatized animals are identified and removed if necessary.

Regardless of the housing configuration and habitat to be established for a species, a larger ecology exists that impacts primates with considerable force. This ecology consists of the surrounding room, its inhabitants including personnel, and the lighting, heating, and ventilation system. Managers of primate colonies are well aware of the influence of an animal across the aisle. Not only are hierarchies established within conspecific groups, they also become established through behavioral expressions, sounds, and odors with animals in near-by cages. Removal of an overly dominant individual from a pair-housed group to the cage across the aisle will do little to reduce the tensions in the submissive individual. An animal temporarily removed from a social group for minor medical treatment should be placed as close to the group as possible in order to reduce trauma associated with re-establishing the social hierarchy. Similarly, return of one or a pair of animals to cages following cleaning, should take into account the social and ecological balance of the original cage location and not replace them indiscriminately within the room. An understanding of this concept is important in order to reduce unnecessary behavioral or physical stress on animals, and to enhance well-being through the application of ecological principles conducive to the well-being of the species.

Personnel, particularly technicians known to the animals, play an important role in the ecology of the room environment. Well-trained, thoughtful persons can be a source of comfort and stability to the room, and should serve the role of the "alpha" individual within the room. Similarly, thoughtless or ill-trained individuals can be a constant source of distress to the animals and a source of considerable variability to the studies to which the animals are assigned.

Additional social propensities that demand attention include those associated with mating, parenting, and playing. Each of these species-typical behaviors requires the attention of knowledgeable personnel in order to ensure their appropriate expression.

Reproduction

Reproduction is both the goal of a breeding program and one measure of the adequacy of the habitat and ecology in which the animals are held. Reproduction is often cited as the necessary indicator of well-being and the adequacy of the cage or environment. Such is most likely far from the truth. Evidence would suggest that some behaviors, such as mating and reproduction, taking care of infants, eating, and elimination are so robust as to be expressed in the presence of environmental inadequacies. Assessments of these behavioral parameters alone would lead one to conclude erroneously that the environment does not need improvement or changing. Many such animals are quite likely housed in this manner today. However, when an *improved* environment is provided to other animals as a means of assessing the former one, differences are likely to become obvious. One means of accomplishing this is to provide choices for the animals, and levels of control over some aspects of their environment, in order to build a template of conditions seemingly most suitable for that species. This template, developed by asking the animal to express preferences, combined with physiological, endocrinological, and behavioral observations will provide a reasonable assurance that the situation is acceptable to the animal and that its state of well-being is as close to optimal as possible.

The Cost of Adaptation

The concept of energy expenditure is worthy of elaboration. Homeostatic mechanisms of animals in all environments constantly strive to return the animal to a baseline level of comfort and well-being. When this process succeeds in enabling the animal to adapt, without excessive behavioral or physiological cost, there is little concern about the animal or the environment. When it becomes evident, however, that this adaptation is only achieved at a great cost, intervention is warranted. Evidence of this occurring cannot be gained by casual or uninformed observation, however, for often the animal raises no red flags, at least in the early stages. The best and earliest warning is through the astute observation of subtle behavioral changes by personnel who know the individual animal. These changes may include such indicators as increased activity, increases or decreases in grooming, shifts in the sleep-wake cycle, altered personality or aggression, or changed eating patterns.

Repetitive motions, often called sterotypic behaviors or stereotypies, are generally considered indicators that the environment is not right for the animal. Although this may be the case, it should also be remembered that these behaviors may be expressions of adaptation to the environment. An example of this is often associated with captive feeding routines. In the wild, primates spend as much as 80% of their time foraging and consuming food. In captivity, this period is frequently condensed to less than 10%. What does the animal do with the rest of the time? It may pace, sleep, groom, or engage in other time-occupying behaviors. Because so much time is applied to these behaviors they may be thought of as maladaptive stereotypies, when in fact they may be very efficient strategies for coping with a new time budget. Much work needs to be done in this area before conclusive recommendations can be developed. In the mean time, careful ongoing observations and interpretations by people knowledgeable about the species and the individual animal are imperative.

Physiological and Biochemical measures

It is sometimes possible to detect changes in pituitary-adrenal hormones during the acute stages of pain or stress, but as stress becomes chronic, biochemical markers become much more variable, and may demonstrate increases, decreases, or no change. Pituitary hormones are often the most reliable indicators of acute pain and stress, and may be expected to include a decrease in levels of insulin, in combination with increased levels of prolactin, growth hormone, glucagon, and epinephrine. This combination of changes contribute to the development of hyperglycemia, which may be utilized in an assessment profile of the animals. The sympathoadrenomedullary system may also respond to pain and stress by increasing plasma levels of catecholamines. The physiological consequences of activation of these systems includes increased cardiac output, increased skeletal muscle blood flow, cutaneous vasoconstriction, reduced gut motility, and increased glucose availability.

It should not be assumed from this brief accounting that a few physiological and biochemical measures can be utilized as reliable indicators of stress or well-being. They can only assist in consideration of many other behavioral and environmental observations.

Maladaptive behaviors

At some point, either through direct observation or by professional judgement, a behavior may be judged to be maladaptive and no longer serve a useful purpose to the animal. Obvious behaviors of this type include self-mutilation, infanticide, excessive grooming leading to hair pulling, aggression, and excessive submission. When it is determined that these behaviors are not caused by pain or disease, predisposing environmental and developmental causes must be examined. It is seldom possible to alleviate these manifestations by pharmacological means, although tranquilizers are sometime effective in helping the animal cope with a change in environment, instituted to achieve a modification in behavior.

Changing cage size and complexity are frequently the first considerations for resolving problems brought on by maladaptive behaviors. Cage sizes are generally selected to enable animals to express species-typical behaviors. While this practice may be commendable and indicative of a high level of concern for the animal's well-being, little evidence is available to suggest that the elaboration of species-typical behaviors are, in themselves, *necessarily*, required for well-being. Also, once maladaptive behaviors become elaborated they are not likely to respond satisfactorily to only a change in cage size. Cage size, per se, does not seem to be the major contributor to maladaptive behaviors. This has been demonstrated in too many colonies in which infant primates were reared without parents, in cages amply large for the development of normal behaviors. In this case, and in many other cases of maladaptive behaviors, the root of the problem lies in lack of adequate social interaction during critical periods of development. Thus, problems may not be due to present-day conditions at all, but may have sprung from social deficiencies at a much earlier age.

This example provides one bit of data that can be used today. When rearing conditions are known to eventually produce maladaptive behaviors in a substantial number of animals at a later age, they should be modified immediately even though the present population of animals is apparently normal.

Importance of Adaptation to Research Data

Many animals may appear to be adapting to colony housing satisfactorily, without showing alarming behavioral or physiological changes. However, even these animals are under stress, the biochemical, cellular, and physiological manifestations of which may profoundly affect some research data. Whether animals are placed into new breeding colonies, or into a colony awaiting assignment to a study, ample time must be provided for this adaptation period. It seems particularly important that periods of adaptation be provided for animals to be studied under conditions of restraint.

Summary

Environmental factors affecting the well-being of primates in captive habitats begins with an institutional commitment to the well-being of the animals, and the training of personnel who will use and care for them. A plan of action is then developed, which details who is ultimately responsible for the animals, who is delegated authority to do what, and a statement of the overall goal. The well-being of nonhuman primates cannot be ensured if these initial steps are not taken. Following this, housing types and configurations must be determined based upon a knowledge of the species and the purposes of the program. Strategies must be developed for developing social groups, and for removing offspring or individual animals from them. Personnel, must be selected and specifically trained to know the species involved, and given the time to learn individual differences so that behaviors of individuals become so well known that subtle changes are readily apparent. Each person should know the extent of their limitations and when to seek assistance from others. The advice of outside specialists should be sought routinely. The well-being and reproductive performance of nonhuman primates can only be ensured by having knowledgeable people who are motivated to provide for the needs of the animal. Many different housing types and husbandry strategies can provide for these needs, but only the right people can ensure that they work.

The Conservation of Nonhuman Primates and its Importance to Public Health

Robert A. Whitney, Jr.

Director, National Center for Research Resources
National Institutes of Health, U.S.A.

Nonhuman primates are indispensable in modern biomedical research, biologics production, and the testing of compounds for toxicity. They are especially valued because of their evolutionary kinship to humans, not only in gross anatomical resemblance and behavior, but in specific physiological and biochemical similarities. Because of this close relationship they offer particular insight into parallel situations in man. Not only were nonhuman primates the key to development of poliomyelitis vaccine, but they also have contributed greatly to our knowledge and understanding of other diseases as malaria, yellow fever, measles, tuberculosis, mental disorders, and viral oncogenesis, to name just a few.

An enormous number of important biomedical research projects in primates are now under way in laboratories throughout the world. In addition, the actual application of the fruits of research depends, to a large extent, on the use of nonhuman primates. Without preliminary testing in these animals, the risks may be too great to apply theoretical knowledge to humans.

The sudden and rapid spread of the acquired immunodeficiency syndrome (AIDS) has shown even more clearly how important nonhuman primates are in biomedical research. Again it is their close resemblance to humans that makes them such valuable models. As you know, the human immunodeficiency virus (HIV) is almost certainly a mutated simian virus. Several simian retroviruses producing AIDS-like diseases in some species of monkey and no disease in others play a central role in drug and vaccine development for HIV. Chimpanzees, susceptible to infection with the human AIDS virus though without clinical disease symptoms, are crucial to testing of potential human AIDS vaccines, just as they were and are essential in hepatitis B research.

The AIDS outbreak is devastating in itself, but it is also a potent reminder that we do not know what public health problems await us in the future for which nonhuman primate models may be essential to our response.

When biomedical research began its great surge of development in the age of Pasteur and Koch, much of the work was done on animals, and significant advances were frequently a direct result of animal observations. Naturally, at that time, the emphasis was on the more readily available animals, however, even then some work with nonhuman primates occurred. Dr. Lebert of Breslau may have been the first investigator to use monkeys as subjects for studies in the transmission of disease. In 1869 he began pathological descriptions of tuberculosis found in monkeys that had died at the Breslau Zoological Gardens.

During the first half of the twnetieth century, primates slowly came into their own as laboratory animal models. The rhesus monkey (Macaca mulatta) became the most commonly used, apparently because it was relatively easy to obtain, a convenient size for most studies, and a relatively hardy animal. The special characteristics of certain species make them particularly valuable for specific studies. This is especially so of the New World (neotropical) monkeys. Findings related to viral oncogenesis in primates involved several of these. Research on drug-resistant strains of malaria depends on the use of owl monkeys (Aotus spp.), which are susceptible to human strains of malaria. The mystax tamarin (Saguinus mystax) is susceptible to human viral hepatitis A, and is an important tool in the development of both diagnostic procedures and a vaccine.

Until the 1970s, almost all of the primates needed for biomedical research were taken from the wild. A study conducted in 1978, as serious shortages of primates were developing, showed that about 34,000 nonhuman primates fron 35 different species were needed yearly by the U.S. biomedical community. More than two-thirds were Old World primates, and half of these were macaques. African green monkeys and baboons made up most of the remainder of the Old World species.

Of the New World monkeys, the identified need was highest for squirrel monkeys (Saimiri sciureus), followed by tamarins (Saguinus spp.), owl monkeys (Aotus spp.), and common marmosets (Callithrix jacchus). Many other species were needed in much smaller quantities for specific research projects.

Owl monkeys and tamarins are especially valuable for research on specific major diseases. Squirrel monkeys are versatile animal models usable in a wide variety of studies. So it is not surprising that these genera remain among the most used New World primates in biomedical research. Owl monkeys, invaluable in malaria research, are also used in glaucoma and vision studies, and in research on respiratory syncytial virus in bronchopneumonia and bronchitis. Tamarins, invaluable for hepatitis A research, are also important in studies of other diseases, especially in immunology, virology, and oncology. The cotton-top tamarin (Saguinus oedipus) is important in research on colon cancer and colitis. The animals used are now all from breeding colonies, since this endangered species is not available from the wild. Squirrel monkeys are valued in areas as varied as behavioral research; malaria, heart disease, and reproduction research; drug development and testing; and space flight research. And it remains true that smaller numbers of other species are needed precisely because they are good models for particular studies.

When the 1978 study was done, destruction of primate habitats, overcropping for the pet trade, and subsistance hunting was threatening the existence of many primate populations, as continues today. Although the loss of primates to the demands of research was and is negligible in comparison with this enormous destruction, the measures taken by the governments of source countries to protect primate populations inevitably affected biomedical research as well. In 1973, India reduced exportation of rhesus by 50%, and in 1978 imposed a total ban on primate exports. The monkey most commonly used in laboratories, the rhesus, became virtually unavailable. Brazil, Colombia, Peru, and other Latin American countries, concerned for conservation, also embargoed exportation of primates.

Of the various programs initiated to protect the international public health by ensuring a continuing supply of nonhuman primates for biomedical research use, I will focus on the initiatives based on the United States National Primate Plan. The 1978 study I mentioned earlier was the starting point for this plan. The plan was developed by the Interagency Primate Steering Committee, chartered by the Assistant Secretary for Health, who heads the United States Public Health Service. The committee's members represented various U.S. government agencies responsible for public health and related activities. The Interagency Primate Steering Committee continue today with a broader mission and a new name: the Interagency Research Animal Committee. I serve as its chairman. Because of the special needs for nonhuman primates in biomedical research and testing, the continuing availability of a sufficient supply remains one of our principal concerns.

The three principal steps of the National Primate Plan remain the fundamental requirements for conserving nonhuman primates needed for the public health, under whatever sponsorship. These three steps are:

- expanding production of primates in the countries using them;
- developing international programs for the production and conservation of several species of primates; and
- making the most effective use of available primates.

In the United States the greatest expansion of domestic production has naturally been that of rhesus monkeys, almost unavailable otherwise. Domestic breeding has other advantages in addition to meeting shortages. The animals should be of higher quality and better standardized. Specific pathogen-free monkey colonies, now rare, will increase, heightening even more these animals' value in viral diease research. The NIH National Center for Research Resources is currently developing specific pathogen-free rhesus monkey breeding colonies to provide better models for use in AIDS research.

Commercial dealers have not developed New World breeding colonies in the United States because of the smaller numbers of these animals required for research. Domestic breeding occurs in the research community, such as the owl monkey breeding colonies at Battelle Northwest Laboratories, the New England Regional Primate Research Center, the Walter Reed Army Institute of Research, and the Bolivian squirrel monkey breeding and research resource colony at the University of South Alabama in Mobile, Alabama.

At this last-named resource, research has concentrated on enlarging our knowledge of the biology of the squirrel monkey. There are two goals for this research: first, to maintain squirrel monkeys better in captivity so we can be protected against losing them through habitat destruction, and second, to make our biomedical research efforts with squirrel monkeys more efficient and fruitful, so that fewer will be needed.

On the second point, development of international production and conservation programs, the most important initiative based on the plan was clearly the PAHO Program of Technical Cooperation for the Conservation, Reproduction, and Utilization of Nonhuman Primates. It is the subject of the next presentation today. I will merely comment that the establishment and development of the Peruvian National Primate Project, including the breeding station at Iquitos, was an immeasurably important move both for the conservation of the primate species involved and for world public health, especially as regards infectious diseases.

The Pan American Health Organization was obviously the proper and effective focal point for inaugurating these international cooperative endeavors so essential to international public health. PAHO's role remains essential in meeting the challenge of the public health needs for nonhuman primates as habitat destruction continues.

The third element, action to assure the most effective use of available primates, has many facets. Among them are not using primates when another kind of animal could produce comparable results, not using a primate species in short supply if a more plentiful species would be adequate, using the minimum number of primates that will produce acceptable scientific results, and human and scientifically compatible recycling and multiple use.

Important as step 1 and step 3 are, domestic breeding colonies and effective use of animals already available will not provide enough primates for the world's public health needs in biomedical research, biologics production, and toxicity testing. Conservation and management of primate species in their native habitats are and will remain important.

It is likely that increased demands will be made for species not widely used now, as new technologies show their usefulness. We do not know and cannot know what primate models will appear in the future, especially in immunology. The recent discovery of a retrovirus in African green monkeys that is harmless to them but deadly to other Old World monkeys can have many parallels in species not now used at all in research.

In light of the accelerating destruction of nonhuman primate habitats since the U.S. National Primate Plan was established, the three-pronged approach spelled out in that plan is all the more important now. For the sake of all humankind, we must press on with domestic breeding colonies, international efforts to continue availability of primates from source countries, and thoughtful, effective use of these irreplaceable resources.

PAHO/WHO Technical Cooperation in the Conservation and Utilization of Non Human Primates in the Americas

Primo V. Arámbulo III

Alfonso Ruíz

PAN AMERICAN HEALTH ORGANIZATION
Regional Office of the World Health Organization
Washington, D. C. 20037

Changes in human distribution and mobilization have brought on destruction of natural habitats of nonhuman primate populations resulting in serious depletion. As a result, there is a growing interest in the development of programs designed to conserve wild primates as a renewable natural resources which may be managed and utilized for the mutual benefit of both man and animals. Nonhuman primates play an important role in public health. They are a key factor in the development of vaccine and they have made great contribution to knowledge and understanding of numerous diseases. The Pan American Health Organization, Regional Office of the World Health Organization for the Americas, recognizes the essential role of nonhuman primates in its overall program of technical cooperation in public health. Through its Veterinary Public Health Program, PAHO is actively collaborating with its Member Governments in developing and strengthening activities for the conservation, reproduction, and rational use of nonhuman primates.

The rapid expansion of the human population mobilizing settlements into the virgin natural areas and the agro-industrial development of several countries, have been the major causes of forest destruction, resulting in depletion of wild primate population and a reduced number of indigenous species.

The increasing awareness of a valuable role of nonhuman primates in public health has created a response of government of several countries to restrict the trade in primates, sometimes by prohibiting their exportation. However this action has not yet solved the basic problem of habitat destruction. Unfortunately, few countries have shown interest in developing programs to preserve wild primates as a renewable natural resources for biomedical use.

As a result of this restlessness, the member countries of the World Health Organization collectively, during the XXVIII and XXIX World Health Assemblies declared their concern about the risk of disappearance of the non-human primates as a result of destruction of their natural habitats. On these assemblies, were approved resolution WHA28.83 and WHA29.67 respectively, by which request WHO Director General to: "Facilitate the exchange of both resources and technology between all countries concerned and, with the help of other interested international organizations as appropriate, to make expert advice available to the countries, at their request, on the conservation, breeding, supply and utilization of non-human primates".

Through this mandate it was expected that the countries would enable national programs for the following purposes:

- To give special attention for conservation in order that primates may thrive in their natural habitat on a permanent basis;
- To manage natural primate population as a renewable resources;
- To urge those member government possessing natural primate populations to control cropping, whether from periodic capture in the wild or from local breeding colonies;
- To cause a positive impact in the community development within designated areas of conservation.

Taking under consideration the above purposes, the following strategies have been set for developing programs of conservation, reproduction and rational utilization of nonhuman primates in the Americas.

- Conservation and breeding in their natural habitats.
- Reproduction in captivity.
- Cooperation among countries in the development and strengthening of national primatology programs.
- Development of appropriate technology to promote the use of indigenous resources.

The main question for the development of national and regional nonhuman primate programs was the dilemma of the two, antagonistic aspects involved:

1. The conservation of primates.
2. The demand of primates for biomedical uses.

In June 1975 the Pan American Health Organization (PAHO) and the Government of Peru convened the "I Interamerican Conference on Conservation and use of Neotropical Nonhuman Primates for Biomedical Research". This conference recommended the establishment of an international group, including PAHO and the International Union for Conservation of Nature and Natural Resources (IUCN) to study the priorities for the development of colonies for primates reproduction in captivity, establishment of natural reservation areas to protect the native species and to develop the taxonomic characterization of the primate species. It was also recommended that all the countries confirm the "Convention on International Trade of Endangered Species on Risk of Extinction (CITES)".

In the mid-1970's PAHO supported the development of national primatology programs in cooperation with the governments of Brazil, Bolivia, Colombia, and Peru. Progress in Colombia was interrupted by the disaster of November 1985 which demolished the Armero Breeding Station, killing almost the entire *Aotus* colony and the responsible investigator. In Brazil the main effort was aimed at training for breeding in captivity at the Ministry of Health's National Primate Center in Belém. The Primatology Program in Peru, set up in 1976, with headquarters at the Primate Breeding and Conservation Center in Iquitos, has evolved into an effective organization that combines management in natural areas with breeding in captivity.

In 1976 the Government of Peru promulgated a national policy for the investigation and management of wildlife fauna, which considered the Primate Project as one of its lines of action. Then, the Pan American Health Organization (PAHO) provided technical cooperation for the implementation and development of Peruvian Primatology Project, initiating census and population dynamic studies to provide scientific bases for controlled cropping and perpetuation of species.

On maturity of the project, PAHO formulated objectives with a wider coverage, as follows:

- To provide recommendations for the protection of wildlife areas suitable to guarantee the protection and conservation of species of nonhuman primates.
- To obtain and provide basic information needed to design and execute management modules for natural population and breeding.
- To develop methods for breeding nonhuman primates in captivity and in their natural habitat, and to promote their conservation in the wild.
- To determine the present status, ecology, distribution and population dynamics of nonhuman primates, especially those needed in biomedical research.
- To obtain information and develop methods to stabilize or upgrade the habitat for selected species of flora and fauna in the forest.
- To collaborate with national authorities in developing nonhuman primates conservation and reproduction plans, and assist in the mobilization of external financial resources for their execution.
- To promote the utilization of primates in program activities and resources as a vehicle for promoting integrated community development in program areas.

The operation of the project required a well defined work-frame of coordinated activities to encompass the interest of the primate user countries with those of the countries developing program for conservation/reproduction.

Figure N° 1 shows the relevant pathways of the global technical cooperation devised by PAHO for the conservation and reproduction of nonhuman primates projects providing both, the political and technical skills for those countries with potentiality of conservation/reproduction programs as well as to those countries requiring primates for biomedical use, by facilitating transfer of primates with recognized origin and to managing income transfer fees to support conservation/reproduction activities.

Figure N° 1 SCHEMATIC DIAGRAM OF PAHO/WHO TECHNICAL COOPERATION FOR CONSERVATION AND REPRODUCTION OF NON HUMAN PRIMATES

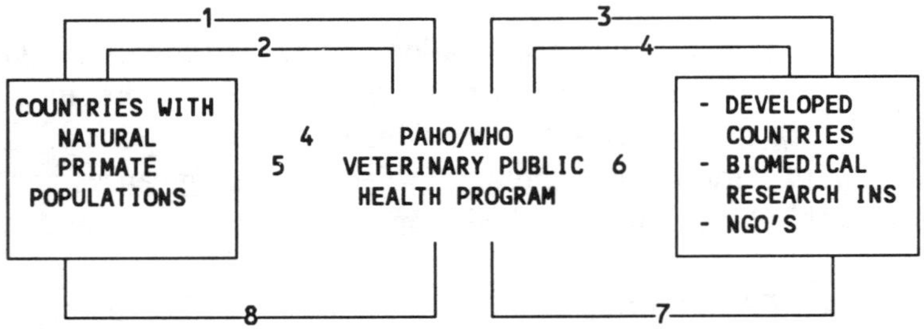

1. Adapts national policy of sustainable development.
2. Cooperates in preparation of strategies/plan of action National Primatology Program.
3. Informs of research programs and primate needs.
4. Mobilizes resources to develop conservation/reproduction programs in countries of origin.
5. Official resolution advises availability primates in excess natural population or born in captivity.
6. Facilitates transfer of primates.
7. Provides transfer/conservation fees, and reports results research progress.
8. Manages income transfer fees to support conservation/reproduction activities.

Implementation of the conservation/reproduction project in the countries require specific structural organization which consists primarily of a National Committee and a Technical Committee. Both are composed by representatives of all the institutions involved in the coordination, implementation and evaluation of programmed activities. The execution of activities are made through Centers for conservation and reproduction of non-human primates. (CRCP), which have three programatic: Breading and Health, Laboratory, Natural areas.

In the case of the Peruvian Primatology Project (PPP), the Ministry of Agriculture, the Ministry of Health, the National Major University of San Marcos, the Research Institute of the Peruvian Amazon and the Pan American Health Organization compose the *Advisory Council* which constitutes the National Committee and share the responsibility of the project (Figure N° 2 and 3).

Figure N° 2. PERUVIAN PRIMATOLOGY PROJECT ORGANITATIONAL CHART

ADVISORY COUNCIL

BOARD OF DIRECTORS

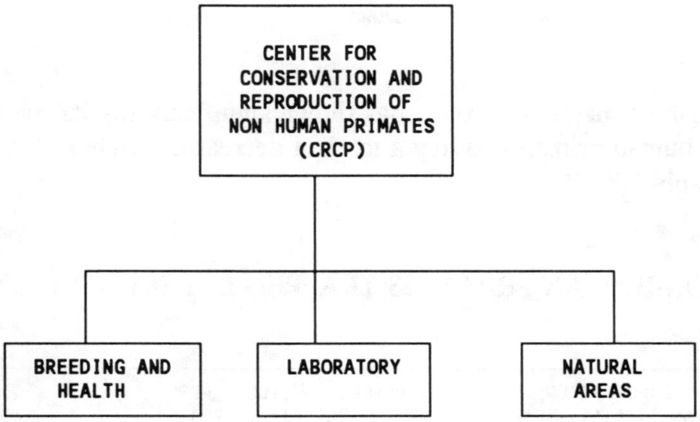

Figure N° 3. PERUVIAN PRIMATOLOGY PROJECT CORPORATION OF ADVISORY COUNCIL

PRESIDENT:	-	MINISTRY OF AGRICULTURE
MEMBERS:	-	MINISTRY OF HEALTH
	-	AMAZON REGION ASSEMBLY
	-	UNIVERSITY OF SAN MARCOS (VITAR)
	-	RESEARCH INSTITUTE OF THE PERUVIAN AMAZON
	-	PAHO/WHO
SECRETARY EX-OFFICIO:	-	EXECUTIVE DIRECTOR, CRCP

The Technical Committee of PPP consist of a Board of Director presided by the Agricultural Division of the Amazon region and constituted by representatives of Health Division of Amazon region, the Research Institute of the Peruvian Amazon, the Veterinary Institute for Tropical and Altitude Research (VITAR) and PAHO/WHO. (Figure N° 4).

Figure N° 4. PERUVIAN PRIMATOLOGY PROJECT COMPOSITION OF BOARD OF DIRECTORS

PRESIDENT:	-	AGRICULTURAL DIVISION AMAZON REGION
MEMBERS:	-	HEALTH DIVISION AMAZON REGION
	-	RESEARCH INSTITUTE OF THE PERUVIAN AMAZON
	-	VETERINARY INSTITUTE FOR TROPICAL AND ALTITUDE RESEARCH (VITAR)
	-	PAHO/WHO

The Peruvian Primatology Project has shown significant results, particularly on the rational explotation of non-human primates having a marked decreased number of exported animals, as it is demonstrated in Table N° 1.

Table N° 1. NON-HUMAN PRIMATES TRAPPED IN PERU 1964-1989

EXPORTATION 1964-1974 (11 YEARS)		TRANSFERS IQUITOS - PERU 1975 - 1989 (14 YEARS)	%
No. Animals	339,080	8,164	
X/Year	30,825	583	
			Decrease 97.6

By means of the Pan American Health Organization and PPP, on estimated number of 22,661 primates are being saved of exportation every year in relation to previous numbers showed during period 1964-1974.

Under the project execution, a total of 8164 primates were transferred to the Pan American Health Organization for further distribution to Biomedical Research Institutes. The main species being transferred are *Saimiri sp.* (34.2%), *Saguinus mystax* (33.3%) and *Aotus sp* (20.4%). Table N° 2.

Table Nº 2 NON-HUMAN PRIMATES TRANSFERED TO PAN AMERICAN HEALTH ORGANIZATION 1976-1989 CENTER FOR BREEDING AND CONSERVATION OF NON-HUMAN PRIMATES (CRCP), PERUVIAN PRIMATOLOGICAL PROJECT (PPP), IQUITOS-PERU, JULY 1990

SPECIES	TOTAL	%
SAGUINUS MYSTAX	2,715	(33.3)
SAGUINUS FUSCICOLIS	661	(8.1)
SAGUINUS LABIATUS	155	(1.9)
SAIMIRI	2,793	(34.2)
CEBUELLA PYGMAEA	150	(1.8)
AOTUS	1,667	(20.4)
CEBUS APELLA	23	(0.3)
TOTAL	8,164	(100.0)

The Center for breeding and conservation of non-human primates in Iquitos has made significant progress, getting an actual breeding population which number 751 distributed in eight different species: *A. nacymai*, *A. vociferans*, *C. pigmea*, *S. fuscicollis*, *S. labiatus*, *S. mystax*, *Saimiri boliviensis* and *S. sciureus* (Table Nº 3).

Table N° 3. PRIMATE BREEDING POPULATION IN THE CENTER FOR BREEDING AND CONSERVATION OF NON-HUMAN PRIMATES, PERUVIAN PRIMATOLOGICAL PROJECT (PPP), IQUITOS - PERU, 1990

SPECIES	ADULT		JUVENILE		OFFSPRING		TOTAL (%)
	M	F	M	F	M	F	
AOTUS							
A. NANCYMAI	82	82	33	38	22	18	275 (36.6)
A. VOCIFERANS	52	60	8	4	4	4	132 (17.6)
C. PIGMAEA	2	3	1	-	-	-	6 (0.8)
SAGUINUS							
S. FUSCICOLLIS	11	8	1	3	-	-	23 (3.1)
S. LABIATUS	7	5	3	3	-	-	18 (2.4)
S. MYSTAX	21	17	4	6	2	2	52 (6.9)
SAIMIRI							
S. BOLIVIENSIS	13	64	26	2	3	12	120 (16.0)
S. SCIUREUS	9	41	32	28	7	8	125 (16.6)
TOTAL	**197**	**280**	**108**	**84**	**38**	**44**	**751 (100.0)**

M = MALE F = FEMALE

Other non evident benefits have been obtain through the development of the Peruvian Primatology Project, particularly in relation to the development of the community settled around the Center activities. This has contributed to the primary health care of the population, community education, intersectoral linkage for eco-development and globally to the improvement of the social and economic well being of the sylvatic area human population.

Wildlife Management in Brazil

Jordan Wallauer Paulo

Medico Veterinário,
Divisão de Fauna e Flora Silvestre,
Instituto Brasileiro do Meio Ambiente e dos Recursos Naturais
Renováveis - IBAMA.
Brasil

The following topics will be discussed:

a) The situation of wildlife management in Brazil, with special emphasis on its historical evolution.

b) Wildlife management planning with a view to Brazil's current situation.

c) Wildlife management trends in Brazil.

d) Current situation of neotropical primate breeding in Brazil.

History

Throughout the entire Brazilian colonization period renewable natural resources, including wildlife, were exploited as though they were inexhaustible, which is both cause and consequence of the cultural heritage still prevailing today. Wildlife is still being used both as a source of food and as pets, which prevails not only in regions of difficult access, but also in the large metropolitan areas, where trading in wildlife is both active and sizable. Today, it is one of the principal factors responsible for the threat of extinction looming over some species, among them most of the primate species of the Atlantic Jungle.

The main reason for the disappearance of wildlife from the Brazilian coast, particularly the northeastern region, was the change caused in the environment during the colonization period by sugarcane and coffee crops. Even today, that phenomenon is still responsible for the pressure exerted on forest and animal resources in the midwestern region (soybeans, corn, and rice) as well as alongside rivers and the main transport routes of the Amazon Region.

Moreover, it should be borne in mind that the growth of agricultural expansion over the last 40 years can be equated to that achieved over the entire 450 previous years, i.e., since the discovery of Brazil. At fault are the technologies used in deforestation and cultivation of the soil.

Hunting was originally used only as a means for acquiring food and still persists in rural areas today. However, with the passing of time, it also became a sport, which, from the 1930s to 1960s, was regulated by norms entirely devoid of any scientific justification. This explains the fact that species considered fit for hunting included Psittacidae, primates, and others that made the sport more interesting.

The almost complete disappearance of some species and the growing pressures exerted on others created an awareness with regard to conservation and thus were the determining factors for the promulgation of Law 5.197 in January 1967, which became known as the Wildlife Protection Act. This Act stipulates that all wildlife shall be the property of the State and that the utilization, pursuit, hunting, or capture thereof is prohibited. The Act furthermore forbids professional hunting, authorizes the trade of wildlife products and subproducts originating from breeding sites, and allows sports hunting under specific conditions.

After the promulgation of Law 5.197, the State began to turn to its enforcement in a very slow and gradual manner, both in terms of the necessary administrative structure and the development and acquisition of technical and scientific knowledge.

The late 1970s saw the emergence of the first regulations defining the breeding sites stipulated by the Law and the necessary conditions were created for their operation. Regulations were restrictive initially, but were relaxed gradually with the incorporation of emerging scientific knowledge, so as to permit the existence of a few wildlife breeding sites for commercial purposes. These regulations also permitted the breeding of wildlife for scientific research purposes, which resulted in the legalization of private wildlife collections with reduced scientific production?.

Sports hunting is practiced only in the State of Rió Grande do Sul, in the extreme southern part of the country, and is regulated on the basis of periodic surveys of the populations of cynegetic species, the setting of game quotas per hunter, and the establishment of annual hunting seasons. Subsistence hunting, especially in the more densely populated areas, is more damaging than sports hunting, because it is practiced year round, i.e., even during the mating season, during which hunting is legally prohibited.

By the mid-1980s, a much more consistent wildlife management model was being used, which took into account cultural traditions, the aemands of environmental activities, and the need for urgent solutions to serious problems. This was a result of broader knowledge both on the dynamics of wildlife species and their actual needs in terms of protection or exploitation potential, and, more importantly, on the technical capacity of the State to understand its own function as administrator of the country's natural resources.

Wildlife Management Planning

It has only been over the last three or four years that the public agencies responsible for policy-making and the administration of wildlife resources have come to understand the need for planning actions in the medium and long term and for responding to the serious problems they face in seeking to satisfy the sometimes contradictory demands of different segments of society. Naturally, directives will be sought, technically more coherent reglamentation will be put forth, and the real problem of wildlife management will be discussed more openly with a view to the country's social and economic aspects. As a result of all this, the Amazonia turtle (*Podocnemis expansa*) now can be bred in Brazil. In fact, it is even possible to breed Pantanal caymans (*Caiman crocodilus yacare*) from this species' eggs, obtained in nature.

These activities were regulated in 1989 and 1990 on the basis of scientific data developed by IBAMA, particularly in the case of the Amazonia turtle, and by a university and a rural entrepreneur, in the case of the caymans.

Studies are still being carried out to find a way of technically improving the situation of lepidopterans, i.e., by increasing their native population through the cultivation of those plants that are part of their diet in small, controlled areas, and by breeding their larvae, as is stipulated by the Law.

This change in behavior exhibited by the public administration has been motivated by a series of factors, some of which are listed below:

1. The development of technical-scientific knowledge and the acquisition of greater experience by personnel charged with the administration of wildlife resources.

2. The existence of overall directives developed by international organizations, such as IUCN, FAO, and the United Nations, which have put forth the following documentation: World Conservation Strategy and the report of the International Commission on Environment and Development (Our Common Future).

3. The participation of entities such as the "World Wildlife Foundation"?, "Conservation International," and others in programs promoting the conservation of Brazilian wildlife.

4. Greater participation of the private sector, which has an interest in the economic possibilities resulting from the sustained use of wildlife.

5. Participation of technicians in training programs and events abroad, as was the case with the "Latin American Seminars for Wildlife Managers", promotion of the U.S. Fish and Wildlife Service, or the courses offered by CATIE in Costa Rica.

However, some difficulties are still encountered by the public administration, which tries to find solutions for problems relating to ecology, flora, and fauna, while, at the same time, attempting to cut public spending. This has translated into significant budget cuts, particularly in the area of personnel expenditures. An attempt is being made at solving these problems by seeking increased participation of the private sector and of nongovernmental organizations.

Particularly with regard to the use of primates in medical research, it can be said that the participation of private laboratories and research institutes is being sought with a view to developing programs and projects promoting the conservation and sustained use of these animals. As an example, mention should be made of two Brazilian hospitals specializing in the treatment of leprosy, which obtained their permit from IBAMA (agency responsible for policy and administration of the environment and renewable natural resources) and developed the breeding of an armadillo species in captivity (*Dasipus novencintus*) with a view to obtaining mitsudina, which is a substance required for the diagnosis of Hansen's Disease.

Trends

The trend followed by wildlife management policies in the short and medium term is that of planning actions in keeping with the directives oriented toward sustainable use of these natural resources and the support of biodiversity. The strategies used are the following:

1. Protection of endangered species, both through the protection of their habitat and the establishment of ex situ genetic banks, as is done with primates of the genus *Leonthopithecus*. This genetic bank is involved in the breeding of these species outside the country, such as in New Jersey, in the United States, and in Great Britain, etc.

2. Development of models for the breeding of animals in captivity based on the handling of eggs and offspring, as is done with the Pantanal cayman and the Amazonia turtle.

3. Incentive for commercially-oriented breeding of species that are not endangered, such as capybara (*Hidrochaerus hydrocheris*), coypus (*Myocastor coypus*), the common pichico (*Callithris jacchus*), etc.

4. Development of studies regarding the viability of hunting certain species that have potential in the area of sports hunting, such as some members of the Tinamidae, Anatidae, and Columbidae species.

5. Authorization for the scientifically-oriented breeding of animals considered endangered, in accordance with Brazilian laws or even in CITES. These nonprofit breeding processes may yield animals for medical research, but should try mainly to contribute to a greater understanding of the biology and the development of techniques for breeding in captivity.

6. Use of surplus animals from zoological gardens, which, in some cases, represent hundreds of species, as in the case of the monkey (Cebus apella).

7. Use of education and extension in changing cultural traditions, thus creating a positive awareness developed around the potential of use of wildlife resources.

As time passes, and depending on the success achieved with the above- mentioned strategies, it can be determined ahead of time that research will require models involving sustained use of wildlife populations, which may be useful in modifying Law 5.197 and may permit the legalization of management involving in situ exploitation, which is currently prohibited. This type of management will certainly be possible only if it goes hand in hand with the creation of an adequate awareness among the beneficiaries at all levels and with the development of control systems.

Breeding of Neotropical Primates in Brazil

The current situation of primate breeding sites is illustrated by the number of registered sites, i.e., ten for scientific purposes and one for commercial purposes. Those sites authorized for scientific breeding produce the following species:

Callithrix aurita
Callithrix jacchus
Callithrix penicillata
Callithrix seoffroyi
Callithris flaviceps
Callithris humeralifer
Callithrix kuhlii
Callithrix argentata
Leontopithecus rosalia rosalia
Leontopithecus r. chrysomellas
Leontopithecus r. chrysopysus
Cebuella pysmea
Sasuinus bicolor
Sasuinus midas
Saimiri sciureus
Callicebus moloch
Chiropotes satanas
Aotus sp. (2 species)
Cebus apella (3 species)
Cebus albifrons
Brachyteles arachnoides

Commercial breeding sites produce the following species: *Callithris jacchus* and *Callithris seoffroyi*, with an exploitation of 60 *C. jacchus* and 14 *C. seoffroyi* in 1990 and an expected exploitation of 120 *C. jacchus* and 40 *C. seoffroyi* in 1991.

Also, although not registered as a breeding site, the Fundacâo Zoological Park of Sâo Paulo does excellent breeding work with three subespecies of *Lontopithecus rosalia*, in conjunction with the Primatology Center of Rió of Janeiro, Zoological Gardens in the United States, and the Jersey Wildlife Preservation Trust, in a program aimed at the preservation of these primates.

Special notice should be taken of experiments performed by the Primatology Center of Rió of Janeiro with hybrids of the *Callithris* species, in search for a litter that is more resistant and fertile and adjusts better to life in captivity, for use in scientific research in general.

However, despite the success achieved in the reproduction of almost all species, particularly those of the genus *Callithrix*, breeding sites established for scientific purposes and associated with public administration are currently having difficulties maintaining their nurseries and expanding their research, as is the case with the Primatology Center of the University of Brasflia, which has practically shut down. Ensuring the continued work of these breedins sites, which, because of their efforts aimed at maintaining biodiversity and their medical research, benefit humanity as a whole, will require cooperation on the part of the private sector and nongovernmental organizations.

Managing Non-human Primates as Renewable Resources: the Economics of Conservation

Theodore Panayotou

Harvard Institute for International Development,
Kennedy School of Government,
Harvard University,
Cambridge, Massachusetts 02138, USA.

Primo Arámbulo III

Veterinary Public Health Program,
Pan American Health Organization,
World Health Organization,
Washington, D.C. 20037, USA.

The ubiquitous presence of non-human primates in zoos has helped increased their appreciation as a source of amusement and education for the general public. Yet their importance in other areas such as public health, biomedical research, ecological balance and the livelihood of local populations continues to be underestimated. Traditionally, non-human primates in the neotropics were trapped as a source of food and cash income. Monkey meat was sold in towns and cities through the Amazon and live animals were actively traded as pets and experimental animals for biomedical research.

In certain areas, such as the Peruvian Amazon, overharvesting and destruction of natural habitats have led to severe depletion of the wild primate population. In response, the Peruvian Government has banned, since 1973, the trapping, holding, and commercialization of non-human primates, while special permits are granted for the trapping of primates for scientific purposes. In an effort to conserve as well as to rationally use the primate population, countries with significant primate population, in collaboration with the Pan American Health Organization, have developed policies and projects of managing non-human primates as a renewable resource. A case in point is the Peruvian Primatological Project which has began in 1972. This project considers non-human primates as a renewable resource which under proper management could provide for the needs of the biomedical research community while at the same time ensuring the long-term survival of theses species in their natural habitats.

The objective of this paper is to review concepts and principles of renewable resource economics potentially relevant to the conservation and rational use of non-human primates. No attempt is made to apply these principles to specific cases or to obtain empirical estimates. We review in order the concepts of maximum sustainable yield, maximum economic yield, opportunity costs of habitats and valuation of approaches for primates.

Maximum sustainable Yield

While each species has its own population dynamics, most biological populations, including non-human primates tend to follow a logistic growth curve. That is, they have low growth at low and high population levels and high growth at some intermediate population levels. This is so because growth is the net result of mortality and reproduction which in turn depend on he size of primate population relative to food availability. At high population levels, growth is also low because both reproduction and mortality are high (due to low food availability per animal). Left undisturbed, a primate population would reach a natural equilibrium at which net population growth is zero (see Figure N° 1). Of course, because of natural fluctuations in food availability (due to irregular fruiting) and in the habitable area (due to flooding), the primate population may vary from year to year even in the absence of disturbance by man due to higher (lower) mortality and/or migration in bad (good) years. such fluctuations may be considerable and should be taken into account in managing primates as a renewable resource.

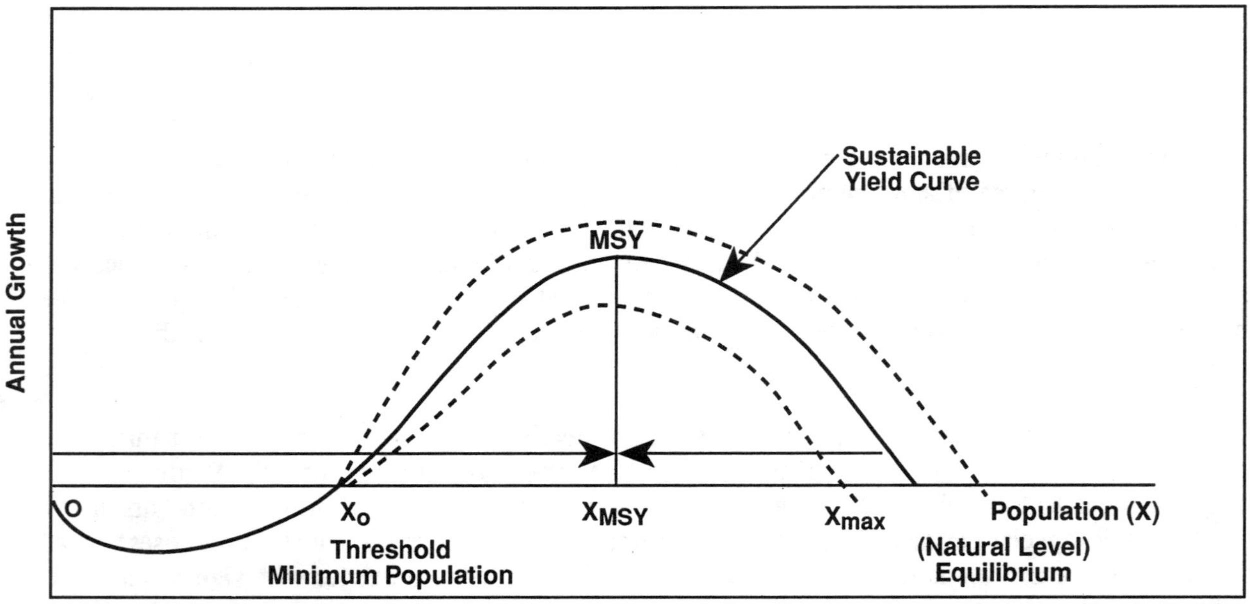

Figure N° 1: Managing primate populations for maximum sustainable yield.

While we can postulate *a priori* that the annual growth is maximized at some intermediate population level where both parental stocks and food availability are relatively high but not excessive, the precise level varies according to species and ecosystem and can only be estimated empirically. In fact, it is more likely to be a range of population levels rather that a single level, because of natural fluctuations in food and habitat. To sustain any given level of population only the annual growth or yield should be harvested, no more and no less. If more is harvested the population would decline; if less is harvested the population would increase. At natural equilibrium of maximum population the growth and therefore the sustainable yield is Zero. To generate growth part of the population must be harvested in the same way that a climax forest must be opened up for net growth to occur. The maximum sustainable yield (MSY) is obtained at the level of population that gives rise to maximum growth. Given past overcropping some primate populations may be at level below that which corresponds to MSY.

Certain species, such as *Cacajao calvus rubicunds* and *lagothrix flavicauda*, are endangered which means that their population has been reduced below the critical minimum level or threshold (Xo); in other words, their numbers have been reduced to levels that do not ensure survival. Others, such as *Cebus albifrons cuscinus* and *aluata palliata* are vulnerable which means that their numbers are approaching the minimum survival levels. These threshold survival levels are specific to individual species and should be conservatively assessed because of natural variability and possible irreversibility, once such thresholds involves the establishment and strict observance of such thresholds, since survival is one of the objectives. of course, since rational use requires maintaining the primate population at the level that generates the maximum sustainable yield, there is a little risk that the population would be reduced below the survival threshold under rational management. However, for species which have a very large minimum survival population and their reproduction rates are low and their food supply is subject to wide fluctuations, maximum sustainable yield management is no guarantee for survival.

Maximum Economic Yield

A biologically more conservative and economically more beneficial management objective is the maximum economic yield (MEY). This objective recognizes that because harvesting and marketing are not costless, it does not pay to harvest the maximum sustainable yield, in the same way that it does not pay to aim for minimum (rather than optimum) output in any production activity. The maximum net revenue or profit is obtained at a lower sustainable yield (growth) and hence higher level of population, than the maximum sustainable yield, as shown in Figure N° 2.

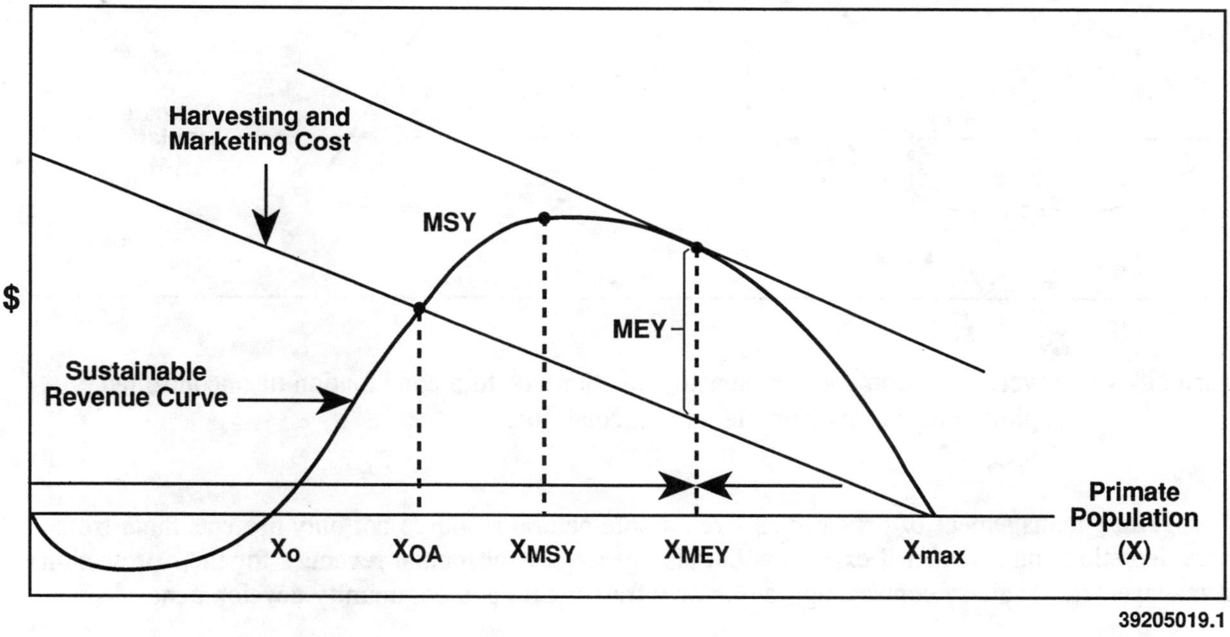

Figure N° 2: Managing primate populations for maximum economic yield.

The sustainable revenue curve of Figure N° 2 is simply the sustainable yield curve of Figure N° 1 multiplied by the price of, say, a primate sold for biomedical research or zoo exhibition. The harvesting and marketing cost curve is the product of the number of primates trapped times the cost of trapping,, transporting and marketing. As seen in Figure N° 2 management for maximum (sustainable) economic yield maintains a significantly larger population of primates in their natural habitat than management for maximum sustainable physical) yield. If. however, uncontrolled exploration takes place under conditions of open access (OA) the primate population would fall to

XOA, and economic yield would be dissipated by excessive harvesting effort. This is the situation that prevailed int he Peruvian Amazon prior to 1973 when the free trapping and trading of primates was banned and permits were granted for controlled cropping for biomedical research.

It is easy to see how uncontrolled harvesting under open access might lead to severe depletion and possible extinction of primates. As shown in figure N° 3, a combination of reduction of the primate population to a low level of XOA and unfavorable natural conditions (heavy flooding, erratic fruiting or disease epidemic) could reduce the primate population below its minimum survival level.

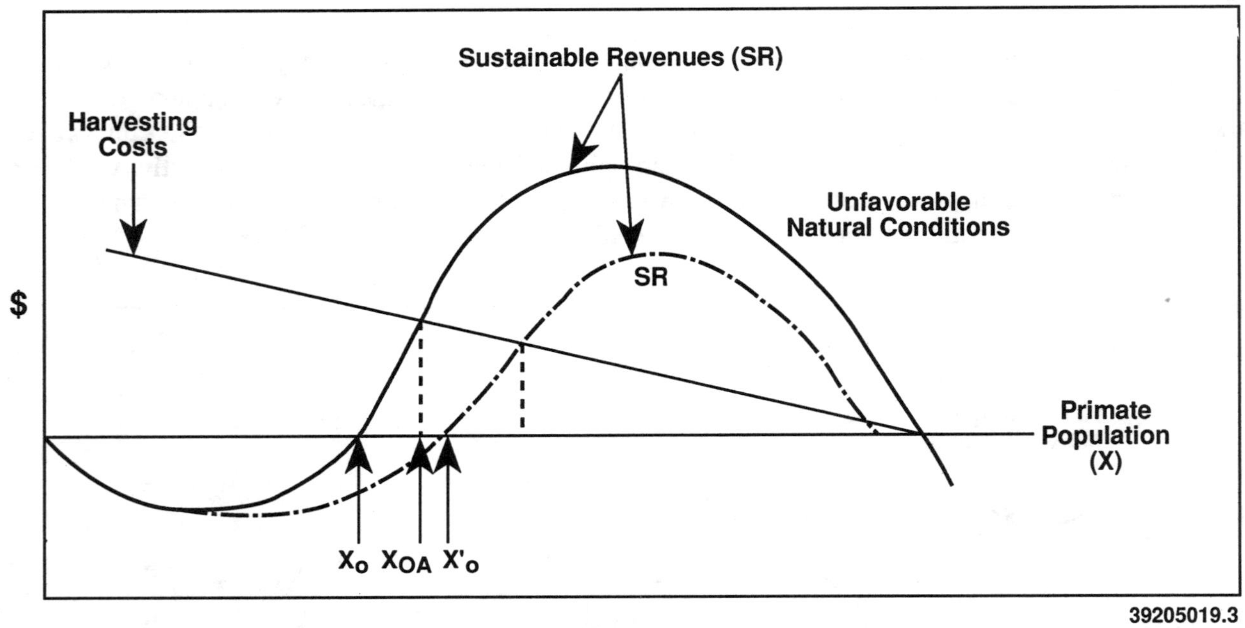

Figure N° 3: Severe depletion and threatened extinction due to a combination of uncontrolled exploitation and unfavorable natural conditions.

Thus, management of primates as a renewable natural resource not only protects them from severe depletion and threatened extinction but also generates substantial revenues for their protection and management, sanitary control, upgrading of infrastructure and community development.

Opportunity Costs of Habitats

If the resources we are dealing with were marine fisheries, the above analysis would have been complete except for the fact that we chose not to discuss the dynamic aspects (at any rate, the maximum dynamic yield is expected to lie somewhere between the MSY and the MEY). However, primates differ in a number of ways from fisheries, whose marine habitat often has no opportunity cost (except in coastal areas), and whose value is usually restricted to revenues from the sale of catch (except in the case of coral reefs where biodiversity values are also involved).

Maintaining a larger population of primates, as it is the case under MEY, management compared to uncontrolled exploitation, requires a larger habitat, that is a larger area of fairly undisturbed forest, as a source of food and living space. if there is a higher-value use for the forest resource (e.g. timber) and for the land (e.g. agriculture), maintaining the area under forest as a habitat for primates has a definite opportunity cost. This is equal to the foregone higher value from timber harvesting and agriculture minus the value of the forest in its natural state (exclusive of the value of the primates). In the case of the Peruvian Primatological Project, we are fortunate to have in approximately the same area a comparative valuation of a hectare of land under 3 alternative uses: ranching, tree plantation, and collection of non-timber products (fruits). The present value of a hectare of land in the Peruvian Amazon near Iquitos was US$9,920 for non-timber forest products, US$3, 184 for timber and pulpwood and US$2,960 for cattle ranching. Therefore, the best use of the land is the maintenance of the forest in its natural state and its exploitation for non-timber forest products. In this case, there is opportunity costs of land or forest in managing primate populations in the same area as a renewable resource for maximum economic yield.

However, to the extent that primates depend for their food supply on the same type of fruits as those that local people depend on for food and cash income there is an opportunity cost to maintaining a larger primate population. Unless alternative sources of livelihood are provided to the local population, managing primates as a renewable resource may proved difficult because people would have a double incentive to hunt the primates: (a) to reduce the competition for fruits, and (b) to supplement their own food supply as well as to earn additional income from the sale of primates. The best approach to address this problem is to involve the local people themselves in the management and controlled trapping of primates for biomedical research. If the local population is effectively given secure rights to the primate resource (subject to the rules of rational management) and technical assistance, they are as likely as anyone to manage the reserve on a sustainable basis.

Value or Primate Resource

When primates were caught and used as a source of meat either for home consumption or sale, their valuation could simply be done on the basis of their market price or the price of a substitute meat. similarly, primates entering the pet trade were easy to value when such trade was legal. With the ban of trapping, holding, and commercialization of primates by Peru and other countries in the region, it is more difficult to obtain market prices for primates sold as pets. However, it is the market expressed aim of governments and PAHO to eliminate such illegal trade. Involvement of the local people in the management and legal trapping of primates is an effective means of doing so, provided that revenues from legal trade exceed those of the illegal trade minus a risk premium, and these benefits are widely distributed to the local population.

The value of primates traded legally for biomedical research and zoo exhibition should be easy to obtain from the governments and or traders involved. There are however two difficulties: (a) the legal market for primates may be too thin for competition to prevail among buyers, and (b) the sources of particular species suitable for biomedical research are few and therefore there is inadequate competition among sellers. Peru and Brazil probably have a duopoly in some species. In terms of husbandry and handling, neotropical primates (concentrated in the Amazon) are easier to manage under laboratory conditions because of their relatively smaller size as compared to their Old World counterparts.

In national parks, wildlife sanctuaries, and nature reserves, primates constitute a part of the wildlife that attracts tourists. The larger the population of primates the more frequent the sighting of primates and therefore the more tourists are likely to visit the particular site. Estimating the viewing value of primates is not an easy task because people visit parks or reserves to view many animals as well as plants, rivers, waterfalls, etc. We need to know the elasticity of tourist visits (and expenditures) with respect to an additional primate (or primate viewing) or to somehow assess people degree of preference (and willingness to pay) between sites with different primate viewing probabilities, keeping everything else constant. In the absence of such a luxury, a creative use of the travel cost and contingent valuation approaches can help obtain some rough estimates. Notice that unlike the previously discussed use values, viewing and photographing primates are not consumptive uses, i.e. they do not diminish the primate population. Adding this value to the consumptive uses discussed earlier would tend at the margin to support a larger population of primates.

Aside from the above use values, primates have several other values which are less quantifiable. First, is their ecological value: primates play a role in the flowering, fructification, dispersal and seed propagation of jungle flora that contributes to plant associations and ecological balance. While it is not possible to place a monetary value on this contribution, its qualitative importance suggests that a somewhat larger population of primates is social optimal than use values alone would justify.

Users of a resource whether engaged in consumptive or non-consumptive use, tend to place an additional value on the resource expressed through options to use the resource in the future as a potential benefit distinct from the actual present use value. This option value is an expression of preference and a willingness to pay to preserve the option of future availability and use. The option value is not limited to current users; non-users may also be willing to pay something to preserve the option of future use should they choose to do so. Thus, biomedical research institutions and scientists, zoos and the viewing public and current and prospective visitors of parks and reserves, would be willing to pay some positive amount to ensure that primates will continue to be available for biomedical experiments, exhibition in zoos, and viewing in parks and reserves. The contingent valuation method is usually used to estimate option values. This is a survey method in which people would be asked what they will be willing to pay for the conservation of primates, or willing to accept as compensation for the loss of primates, or wiling to accept as compensation for the loss of primates. In the absence of actual payments the free-rider problem is solved through proper survey design.

Second, non-human primates have a strategic importance in that they constitute an invaluable genetic pool that should be preserved indefinitely. This is a sort of option value for the society as a whole. It arises because of the uncertainty of future needs and preferences and the irreversibility of extinction. In the case of non-human primates this value is likely to be very high because of their phylogenetic relationship to man. There are major physiological and behavioral similarities, that make primates extremely valuable models for biomedical research. Some have naturally occurring disease conditions similar to man, making them useful in studying human diseases and as experimental subjects in the testing and quality control of vaccines and drugs. This social, or rather global option value is both impossible to estimate and difficult to capture except via international contributions. The assistance of PAHO and NIH to countries with significant primate populations among others are such contributions.

Finally, primates, like other renewable resources and natural environments have an intrinsic value of their own which, though difficult to define let alone measure, is manifested in the many environmental funds and organizations in existence to protect endangered species. People value the existence of a primate, even if they do not plan to use it directly in research, view it, or photograph it (existence value). People also want to assure the existence of primates and other wildlife populations as a bequest to future generations (bequest value). People also have a certain sympathy for animals, including (if not especially) for non-human primates, and are willing to pay something to preserve their habitats and to ensure their survival and well-being. a case in point are efforts by WHO, in response to international concern, to reduce the use of animals in research laboratories, to find alternatives, and to set rules and regulations for the humane treatment of hon-human primates. The existence value can be estimated using the contingent valuation approach as explained earlier.

Conclusion

The greatest threat to non-human primates is the destruction of their habitats by slush-and-burn agricultural activities. Management of hon-human primate populations as renewable resources would achieve the dual objective of conservation and rational use, which re indeed the two sides of the same coin.

For two reasons management for maximum (sustainable) economic yield is preferable to management for maximum sustainable (physical) yield: (a) it maintains a larger population which is an insurance against natural fluctuations and high minimum survival thresholds; and (b) it raises the maximum amount of revenues for the protection of the resources and community development that would provide alternative livelihoods to poaching by local people. The most cost effective means for poaching by local people. The most cost effective means for implementing a renewable resource management policy is to involve and empower the local people in the management of the resources.

The opportunity costs of maintaining a larger population of primates in the neotropics tend to be low while there are significant additional values, such as use values for tourists, ecological values, option and existence values. The optimum primate population level and rate of harvest is an empirical question depending on the population dynamics of the particular species, its use and non-use values and the opportunity costs of habitat and harvest. Managing non-human primates as a renewable resource of significant economic value would conserve the resource for the benefit of both man and animals.

INTEGRATED DEVELOPMENT OF CONSERVATION UNITS AND THE ROLE OF THE NATIONAL PRIMATOLOGY PROGRAMS

Jose Vincente Rodriguez

Instituto Nacional de Los Recuros Naturales (INDERENA)
Renovables y del Ambiente
Bogota, Colombia

With an extension of 1,141,784 Km2, Colombia is located in the intertropical zone at the northwestern corner of South America with coasts both over the Caribbean Sea and the Pacific Ocean and is crossed by the equatorial line. Approximately the eastern half of its territory is composed of lowland rolling plains, belonging to the Amazonian and Orinoco drainages. Mostly in the Amazonia outcrops, tablelands, and mountains appear, which geologically and biologically are related to the Guyana highlands. Furthermore, in the southwestern corner of the Orinoco drainage system, adjacent to the Andes, rises another isolated range, the Cordillera de La Macarena. The western half of the country presents three large branches of the Cordillera de los Andes. The Eastern Andes (Cordillera Oriental), continuous with the Venezuelan Andes, the Central Andes (Cordillera Central) which are a northward prolongation of the Eastern Andes of Ecuador (Cordillera Real), and the Western Andes (Cordillera Occidental). In the Caribbean coast is located the Sierra Nevada de Santa Marta, a mountian massif entirely isolated from the above mentioned Cordilleras, in which the highest summit in the country is located (la Horqueta of Chundua, 5780 m.). Parallel to the northern portion of the Pacific coast is the Serrania del Baudo, and along the Panamanian border is the Serrania del Darien.

The climate is quite varied since it covers a wide range of conditions, for instance, in the Guajira peninsula which is the northernmost tip of South American with an average annual rainfall of 150 mm, warm desert conditions prevail, and in the lowlands of the Choco (Pacific lowlands) the average annual rainfall reaches 13,200 mm. The temperature decreases with increasing elevation so that at about 4,800 m above sea level permanent snow is found. The orographic and climate complexity, together with its geographical situation and a rather complicated geological history, contributes to the development of the fairly complex biotic mosaic and perhaps the highest biodiversity of the world with approximately 10% of the living species of organisms.

Except for Brazil, Colombia has the greatest diversity of primates in the Neotropical region. The three living families of Platyrrhini are represented as follows: *Callithricidae* (2 genera with 8 species); *Callimicomidae* (monotypic family) one species; and *Cebidae* (9 genera with 19 species). In addition there is a record for *Chiropotes satanas* from the Colombian bank of the Orinoco River, which requires confirmation, and in addition at least the following species could be expected to occur in Colombia: *Cacajoa calvus*, *Aotus trivirgatus*, *Pithecia aequatorialis*, *Cebus nigrivittatus*, and *Saguinus labiatus*. Most of these occur in the forests of the lowlands, under 1,000 m above sea level. In the pluvial forest, hygrophitic (rain forest), subhygrophitic (semi-deciduous forest), hygrotrophophytic (deciduous forest) and gallery forest; with the highest diversity in the Amazonian piedmont forest between the Caqueta and the Putumayo rivers. Several species, mainly the medium and large sized one, reach the cloud and Andean forest at altitudes of 2,800 m above sea level (*Cebus apella*), and locally 3,000-3,200 m (*Alouatta seniculus*, *Aotus lemurinus*, and *Lagothrix lagotricha lugens*. Three species are endemic as follows: *Saguinus oedipus*, *Saguinus leucopus*, and *Aotus brumbacki*, and other two yet unnamed species of *Aotus*.

As in several other tropical countries, the greatest problem for wildlife conservation lies in the fact that deforestation has destroyed, by now, more than half of the virgin forest, particularly in the western half of the country, including the Andean region which is heavily populated. The clearing of forests still continues, motivated by haevy pressures toward new lands for agricultural and cattle activities. To these pressures is added that of subsistance hunting. This problem is aggravated by the circumstance that much of the cleared lands are not suitable for sustainable farming or cattle development. Furthermore, large areas with soils of good quality are dedicated to the low-profit extensive cattle industry, and, on the other hand, many of the lands that are heavily used for agricultural purposes have a definitive forest vocation. This prevalent situation continuously endangers a growing proportion of the flora and fauna of Colombia, and calls for urgent and efficient strategies for conservation and protection. In terms of conservation of nonhuman primates indiginous to the country, a progam of conservation should include:

- the creation of new national parks and other reserves, and the consolidation of exisiting ones,
- basic research to complement fundamental biological and ecological knowledge of primate species, their population status, and an adequate monitoring system to watch these populations and the condition of their habitats,
- projects directed toward a better understanding of the breeding biology and other critical aspects, that might facilitate captive breeding projects directed toward providing stock to reestablish natural populations, as well as provide animals for biomedical research,
- educational programs to promote increasingly better attitudes in the human communities toward nonhuman primates,
- and, the development of cooperative mechanisms in terms of research projects, personnel training, provision of equipment and facilities, scientific and technical advice, and financial support.

At present, Colombia has 41 areas in the National Park system, managed by INDERENA, that together with the indigenous peoples reservations, reach over 30 million hectares, of which 9 million belong to different categories of the National Park system. The area of these regions is quite impressive, unfortunately those reserves are not yeat sufficient to provide protection to the 47 species and subspecies of nonhuman primates that occur in Colombia, since new ones should be created and implemented for this aim. Creation of new areas is also important in order to get an appropriate representation of the great variety of ecosystems, endemism centers, and endangered elements of the biota. An analysis of the problems and solutions involved with regard to these topics is presented.

MULTISECTORAL AND INTERCOUNTRY COOPERATION FOR THE DEVELOPMENT OF NATIONAL PRIMATOLOGY PROGRAMS

Manuel Uceda Castillo

Director General of Forestry and Fauna of the Ministry of Agriculture of Peru
Lima, Peru

and

President of the Advisory Council of the Peruvian Primatology Project "Manuel Moro Sommo"
Iquitos, Peru

Gentlemen:

It is a great honor to have the opportunity to speak to you, thanks to the kind invitation of the Pan American Health Organization/World Health Organization, to present to you our proposal on multisectoral and intercountry cooperation for the development of national primatology programs.

Peru, as a member of the World Health Organization, is committed to participating in this meeting since it considers that its contribution will have to constitute an alternative for improving human health and preserving our natural resources and, by that means, the possibility of achieving our development.

Since the end of 1972, because of the world-wide concern for conservation and rational use of nonhuman primates, the Government of Peru, through its Ministries of Agriculture, Health, and Foreign Relations, as well as the Veterinary Institute for Tropical and Altitude Research (VITAR), and with the firm support of the Pan American Health Organization/World Health Organization, initiated the first efforts of research, evaluation, and management of wild fauna in the Peruvian Amazon and tropical forests of the Department of Tumbes on the northern coast of our country, with the responsibility for the orientation and supervision of the execution of the project falling on the Division of Forestry and Fauna of the Ministry of Agriculture.

In 1975 this interest was consolidated through the functioning of the Peruvian Primatology Project "Manuel Moro Sommo" by virtue of the signing of a Letter of Agreement with the Pan American Health Organization/World Health Organization, which after three amendments extended its life until December of the current year. As it is evident, the renewal of the Letter of Agreement is expected after an evaluation of what was accomplished as well as of the characteristics and noble ends which the project exemplifies for the good of the health of the world.

The primary objective of the project is research oriented toward the conservation and management of nonhuman primates in Peru. For its execution three lines of action have been developed to promote the participation of the local population in the conservation of this natural resource: a) evaluation and research; b) conservation and management; and c) education, training, and extension in the area in which these species of mammals are distributed.

We call attention to some achievements and benefits that have been obtained from the project, among them:

- New species of nonhuman primates in Peru have been identified, with a current registration of 32 species, as well as significant information concerning their distribution, the status of their conservation, ecology, ethology, and guidelines for their management and sustained utilization.

- Official establishment of breeding center, the Center for Reproduction and Conservation of Primates, in the city of Iquitos, which has made it possible to obtain greater information and experience in the management, breeding, and reproduction in captivity of some seven species of primates of biomedical use. In addition, it has a specialized library which serves the scientific community.

- Utilization of islands in the Amazon River basin for the introduction of some species of primates, such as *Saguinus mystax* and *Saguinus labiatus*, and their subsequent management in semicaptivity with achievement of notable success in the growth of the populations.

- The understanding and participation of the island communities in the actions for the management of primates in semicaptivity on the islands have been achieved through the protection of the primates; the development of agriculture and silviculture through demonstration plots, for the purpose of halting agricultural migration, and other benefits which the project provides the community in coordination with the other public sectors, such as health, for health care for the local inhabitants; education, through extension and training campaigns; and also support with infrastructure.

- Protection of species of primates threatened with extinction or in vulnerable situations because of clandestine hunting and, principally, the destruction of their habitat, a consequence of the deforestation for agricultural purposes.

Therefore, in addition to Manu National Park, the Pacaya-Samiria National Reserve, and other conservation units in the Selva region, the creation of the communal reserves of Tamshiyacu-Tahuayo (Dept. of Loreto) and of Tahuamanu (Dept. of Madre de Dios) has been proposed, as well as the transfer of primates to reserved areas to guarantee their survival.

- Training of a team of professional and technical personnel at the highest level in the field of primatology, whose contributions are recognized by the international scientific community.

- Contribution with the international scientific research centers, most particularly in the area of biomedical studies, through the transfer of 8,405 specimens of primates selected from the species: *Saguinus mystax* (2,745), *Saguinus fuscicollis* (661), *Saguinus labiatus* (255), *Saimiri* sp. (2,872), *Aotus* sp. (1,699), *Cebuella pygmaea* (150), and *Cebus apella* (23). This has great relevance since our monkeys are excellent models for the production of vaccines against viral hepatitis, malaria, cancer, and AIDS among other ills that seriously affect humanity.

The Peruvian Primatology Project, with almost 18 years of continuous work, constitutes a model for the management of wild fauna; its achievements and benefits are the result of the dedication and coordinated effort of national and international professional and technical institutions and the community as a whole.

The project is developed on the basis of the document "Estrategias y Plan de Acción Nacional 1985-1990 del Projecto Peruano de Primatología" [National Strategies and Plan of Action for the Peruvian Primatology Project for 1985-1990"].

At the regional level, the individual charged with coordinating the actions and responsible for the execution of the project and the implementation of the agreements and policies emanating from the Advisory Council is the Board of Directors, with headquarters in the city of Iquitos.

With respect to the execution of the project, this will be accomplished through its Divisions of Administration and of Research and Management.

Notwithstanding this organization, an effort is made to establish actions coordinated with the education, fishery, interior, and education sectors, among others, as well as with nongovernmental organizations (NGOs) and community organizations.

Finally, some legal provisions were made; however, we should note the reference to the categorization of species of wild fauna according to their conservation status, through R. M. No. 01082-90-AG-DGFF of 14 September 1990 and the Code of the Environment and Natural Resources, through Legislative Decree No. 611 of 7 September 1990.

In parallel with this and other efforts, significant coordination has been accomplished with neighboring countries, with those with whom we share a common heritage; the achievements of such actions are as follows:

- Agreement for the Conservation of the Flora and Fauna of the Amazon Territories of the Republic of Peru and the Federative Republic of Brazil, signed on 7 November 1975 and adopted through Decree No. 21670 of 26 October 1976.

- Agreement for the Conservation of the Flora and Fauna of the Amazon Territories of the Republic of Colombia and the Republic of Peru, signed on 30 March 1979.

- Creation of the Intergovernmental Technical Committee for the Protection and Management of the Amazon Flora and Fauna at the First Technical Meeting on National Parks, Management, and Conservation of the Amazon Biota (Leticia, 28 to 30 August 1975), a result of the initiative of the Government of Colombia during the Meeting of Ministers of Agriculture in Cali in 1975 with representatives of Brazil, Colombia, Ecuador, Peru, Venezuela, and Bolivia in attendance.

The Government of Perú is fully aware that security, sustained utilization, and defense of those resources of wild flora and fauna, in particular the nonhuman primates, which constitute a solid base for the harmonious development of the region, must be carried out by joint action of the countries of America, producers of neotropical primates, with those who are united by the same interest in stable development for our villages and the noble ends of human and animal health. Thus it is with great satisfaction that we view this meeting, in which the officials of the Governments of the region discuss the common actions that they should undertake for the protection and management of the nonhuman primates in particular and of the neotropical fauna and flora in general, resources which we well know have been attacked and which demand the greatest attention in order to contribute to the promotion of the general well-being.

Therefore, as a means of contributing to the discussion which must continue, I propose the following recommendations which, I believe, can constitute adequate channels of cooperation among our countries.

1. The Regional Committee on Primatology for the Americas (CORP) should be formed by the countries of America that possess natural populations of nonhuman primates.

The presidency of the Regional Committee should fall to one of the representatives of the countries possessing the resource, while the secretariat could be the responsibility of the representation of the Pan American Health Organization.

2. For the operation of this Regional Committee, a commission should be formed to prepare and propose a set of rules and regulations for the organization and functioning of this body, to be approved by the General Assembly.

3. The member countries of CORP should commit themselves to utilize only primates coming from breeding centers or natural areas under management, duly established and controlled by the state.

4. Biomedical research in the countries producing primates should promote and support biomedical research in the countries producing primates.

5. The possession, transport, and marketing of primates as pets and their remains should be prohibited and strict control of the traffic in primates should be maintained through reinforcement of the border services and the application of the Convention on International Trade in Endangered Species (CITES).

6. CORP can constitute an excellent permanent channel for information on the advances in research and management and in the national programs or projects on nonhuman primates as well as for the exchange of professionals in the area, and the Center for Reproduction and conservation of Primates in Iquitos could constitute the headquarters of the Center for Research in Primatology in Latin America.

7. The member countries of CORP should take the actions necessary for the establishment of biomedical research laboratories in the countries in which the extraction of primates is carried out.

8. The member countries of CORP should reserve areas for the protection, in their habitat, of species of primates threatened with extinction; recommendation should be made to PAHO/WHO that it request the research centers and laboratories that use neotropical primates to see that the final biological products obtained revert to the benefit of the countries producing the resource.

9. The signing of bilateral agreements between countries producing primates is recommended to strengthen the management of CORP and promote the conservation of the resources of wild flora and fauna.

10. All agreements signed by the Pan American Health Organization that commit the primate resources of a member country should have the written endorsement of the official representative of the respective country.

11. The most appropriate mechanisms for financing the meeting and operation of CORP through PAHO/WHO should be studied.

12. The participation of the public and private sectors should be encouraged, as should that of the nongovernmental agencies, for the establishment of breeding centers and the management in semicaptivity of species that are threatened or that are of high scientific, cultural, and/or commercial value. It would also be advisable to study the legal impediments with respect to the functioning of state programs.

13. The participation of the local inhabitants in the development of the projects and programs in primatology should be promoted so that they are executers and beneficiaries of the contributions that such actions can generate.

14. Efforts directed toward the execution of campaigns for training and extension at every level and by all communications media should not be reduced, with the objective of emphasizing the socioeconomic, scientific, cultural, ecological, and strategic importance of the primates in particular and of the resources of wild flora and fauna in general.

In summary, we representatives to CORP should not forget that the political guidelines that orient the actions for the conservation of the primates should be based on assuring the survival and an increase in the population of the species, on rationally managing the primate populations, and on benefitting the workers in the field, the national professionals, the centers for research and cultural information, and, finally, all the member countries of CORP.

Fellow representatives, this I Reunion of CORP to which we dedicate our efforts is highly significant for the preservation of our resources and a means of improving the quality of life of all the inhabitants of the Region, and, we say, to world health.

Therefore, let me, on behalf of my Government, express my appreciation to the representatives of the member countries of the Regional Committee on Primatology for the Americas, to the Pan American Health Organization/World Health Organization, and to the Government and people of the United States of America for their kind attention.

Thank you.

The Mobilization of Resources for the Conservation of Neotropical Ecosystems: "Debt for Nature" Exchanges

Edward Liebow

Battelle Human Affairs Research Centers
Seattle, Washington USA

Introduction and Overview

Possible approaches to the preservation of neotropical primate populations have three broad sets of aims: (1) habitat conservation; (2) protection of primate populations; and (3) public education. This presentation focuses on a particular approach to habitat conservation, the "Debit-for-Nature Exhange," although it should be recognized that no specific obstacle prevents the application of debt/nature exchanges to programs protecting primate populations or educating the public. First proposed by the president of the US-based World Wildlife Fund in 1984, fifteen such exchanges have occurred to date, 9 of them in Latin America and the Caribbean. Pan-American exchanges have generated more than US $50 million in conservation funds in Costa Rica (US $42.3 million), Ecuador (US $10 million), Bolivia (US $250,000), and the Dominican Republic (US $600,000) since 1987. To date, the main North American groups associated with these exchanges have included the World Wildlife Fund, Conservation International, and the Nature Conservancy. The US government is considering direct support for this sort of exchange, but specific authorization is still forthcoming.

On the surface, this innovative scheme is easy to understand: large amounts of commercial debt are owed to lending institutions in North America, Japan, and Europe, and many of these debts appear unlikely to be paid in full. With the debt/nature exchange, a debtor nation may be forgiven a modest part of its foreign debt and a fund set up through one of a number of financial transactions that generates revenues to support the pursuit of conservation aims. The simplicity of the scheme, however, belies the complex interactions that have contributed to the onslaught of tropical deforestation in the first place. Developing countries with large external debts have consistently attempted to increase exports to gain cash to pay off their loans. For the most part, these efforts have been made at the expense of tropical forests, which have been razed for timber sales or to make way for cash crops. The debt/nature schemes provide incentives to avoid this practice and encourage sustainable development practices while protecting valuable habitat.

The scheme relies on three fundamental principles of action:

- Successful debt/nature programs are possible only where strong coservation institutions exist in the debtor country; and

- To succeed, debt/nature programs must be fair, and must be seen to be fair, that is, the interests of all parties involved must be taken into account.

- From the creditor country's viewpoint, private philanthropic donors and lending institutions are being mobilized, and their continued participation depends upon receiving some assurances that the funds they donate are achieving their maximum effect.

It is to this end -- reassuring donors in creditor countries -- that this presentation recommends some technology transfers that help systematize the consideration of social and economic factors along with biophysical criteria used to select appropriate conservation strategies.

My goal is to briefly describe the mechanics of debt/nature exchange, bearing in mind that every individual transaction is highly distinctive and financially complex. I will then briefly touch on the dilemma formed at the intersection of two technical data issues, habitat values and land tenure. It raises the question: How to strategically target the appropriate lands to set aside for habitat conservation, given that those lands with the greatest habitat value may involve ownership or land use patterns that are difficult to change, while lands that are most readily available for conservation purposes may not have as significant a habitat value?

How It Works: The Five Steps to a Debt/Nature Exchange

Parties to the exchange usually include: (1) The creditor country's taxation and financial regulatory authorities; (2) the creditor country's lending institutions; (3) the creditor country's conservation organization and its financial supporters; (4) the debtor country's government finance (e.g., central bank) and resource management authorities (e.g., agricultural and wildlife ministries); (5) the debtor country's conservation organizations; and (6) local communities potentially affected by conservation programs. Each party has different (and sometimes conflicting) interests at stake in the exchange, so the arrangements for individual exchange programs must be uniquely tailored to take into account the specific interests of all parties. In general, however, these arrangements are achieved in five steps:

1. **Obtain approval in principle from the debtor country for financing mechanisms and conservation program goals.** Along with the creditor country parties to an agreement, three key debtor country parties are involved in negotiations: the government, the cental bank, and an appropriate private conservation organization that will receive the funds and manage the agreed-upon program. The conservation organization may be the most important of the three, in that conservation organizations in developed countries must relinquish effective control over funds to avoid the "imperialism" issue, but must assure donors and tax authorities of the continuing responsible use of funds. Only strong private conservation organizations in the debtor country can meet this dual set of conditions.

Negotiations between these three parties can cover a wide range of issues, including the exchange rate to be applied in converting debt into local currency, the conditions of payment or a bond that may be issued in exchange, and how to design the conservation program to which the proceeds are applied.

2. **Obtaining the debt instrument.** In this step, the conservation organization in the creditor country raises the funds to purchase the debt instrument on the commercial market. There is an established secondary market for debt from developing countries that will set a base price for the debt of the country involved. Market prices vary considerably, ranginf from a few cents on the US

dollar to almost parity. The lower the price, the greater the value for a nature exchange, but market price is also an indicator of the risks involved. The money needed for debt acquisition can be raised in many ways. For organizations in developed countries, this is a classic fund-raising issue, and inevitably brings into consideration whether such a program taps new funding sources or poses a conflict with other fund-raising efforts. The tax status of donations must also be kept in mind.

3. **Transfer of title to the debt.** This is a complex transaction. One important decision is who the actual purchaser will be. It may be appropriate for the conservation organization in the creditor nation to acquire the debt and then donate it to its partner in the debtor country. It may be possible to donate the necessary resources to permit direct acquisition of the debt by the debtor country conservation organization. In light of a recent ruling the the US Department of Treasury, it may also be possible to have the debt donated directly to the conservation organization in the debtor country, acting as an agent for the creditor country organization. Taxes and other financial considerations govern this choice.

4. **Conversion of the debt in accordance with the agreement reached with the debtor country's government.** It can involve the issuance of local currency bonds, measures to protect certain sensitive areas (for example, purchase of lands or protective legislation), and cash payment of local currency in redemption of the debt. Bonds are useful in that they can have an extended maturity date, generating revenue for start-up programs, and, upon maturity, becoming an endowment for future activities.

5. **Execution of the agreed-upon conservation program.** This is the ultimate goal of the program, and needs to be monitored carefully. At the same time that conservation aims are sought, the reputation of the creditor country conservation organization can be at stake, because it must be able to assure donors and tax authorities that the funds it has received are being used appropriately.

Habitat Value and Land Tenure

For debtor nations to derive enviromental benfits from these transactions, the selection of critical habitat areas is key. However, it is equally important to understand local land tenure arrangements in their cultural context, so that managing such areas for habitat conservation purposes avoids a whole cascade of adverse social and economic consequences, leaving debtor governments unwilling to consider such exchanges.

Habitat evaluation is, of course, based on such features as biodiversity, stability, intrusion of nonindigenous species, and sensitivity to disruption of human activities. The distribution of habitat values is often represented on maps identifying areas that require special resource management attention. And even where conservation efforts have been undertaken, most conservation areas (e.g., national parks) represent only part of the specific ecosystem they were intended to protect. The identification of "buffer zones" around valuable park lands can prove helpful in easing pressure on the more encompassing ecosystem within which the parks are contained.

Frequently left out of this identification effort is comparable assessment of land ownership and land uses. Such socioeconomic factors can be especially important where use rights have developed over a lengthy historical period, reflecting larger cultural patterns of patron-tenant relations, expansion at the frontier spurred on by population growth that strains the ability of urban areas to keep up, and publicly sponsored developed schemes to produce cash crops for export. In addition, the land uses, range, and settlement patterns of indigenous peoples may need further documentation, especially in areas being considered as "buffer zones" around conservation areas.

Combining information regarding habitat values and land tenure can be done with the aid of a computer-based Geographic Information System (GIS), which organizes information in "layers" that can be examined separately or in combination with variables contained in other layers. The technology is familiar to conservation organizations in Pan-American countries, but its application to problems of institutional mapping may be lacking at present.

Such a technology transfer would produce at least two benefits. First, candidate habitat areas could be systematically evaluated and given priority rankings based on a combination of social and economic, as well as biophysical factors. The planning basis for sustained, long-term conservation efforts could potentially be reinforced. As will be evident in the Costa Rican case described below, for governments to capitalize on initial successes, they need to have a series of projects lined up and ready to recommend for implementation. The second benefit of encouraging the transfer of geographic information system technology would be that this enhanced analytic capability would be more widely available within the country, with organizations that have heightened sensitivity to local concerns, thus limiting reliance on outside expertise. The projects that would be recommended for implementation would therefore be likely to accurately reflect local interests and priorities, key ingredients for success within the appropriate institutional framework.

Three Case Summaries: Bolivia, Costa Rica, and Ecuador

1. Bolivia

The first debt/nature exchange was essentially one environmental organization's agreement with Bolivian government to help finance and ensure the protective management of more than 1.5 million hectares (4 million acres) of tropical forests and grasslands.

Conservation International, a Washington-based non-profit organization, bought US $650,000 of Bolivia's commercial bank debt, using Citibank as its agent in the secondary financial markets. The debt was purchased at about 85% discount -- that is, 15 cents per US dollar. A $100,000 grant from the Frank Weden Foundation financed this transaction.

Under the terms of the 1987 agreement, Bolivia's obligation to pay the $650,000 debt would be cancelled. In return, the Bolivian government agreed to give the maximum legal protection to the Beni Biosphere Reserve, in the country's northeast, and to increase to 1.2 million hectares (3.7 million acres) the protected areas next to the reserve. The government also agreed to provide an operating fund in local currency worth $250,000 to manage the Beni Reserve and its additional buffer zones for sustainable uses.

The forests and grasslands of the Beni Biosphere Reserve support 13 endangered species and 500 species of birds. The Reserve is also home to the Chimane Indians. The newly protected areas will be managed to use the forest in a sustainable manner, protecting soils and the watershed while protecting traditional Chimane subsistence pursuits.

Conservation International has become a scientific and technical adviser on conservation and resource management to the Bolivian government. To carry out the conservation programs, the organization agreed to designate as its representative a newly formed national commission, subject to the approval of the Ministry of Agriculture.

The national commission includes local officials of the Beni region, university scientists from the Ecology Institute and the Beni Biological Station, as well as a representative of an environmental coalition, Liga de Defensa del Medio Ambiente (LIDEMA). This commission is responsible for carrying out the environmental programs.

The $250,000 operating fund in local currency was created with the US $100,000 equivalent in pesos from the Bolivian government and $150,000 in pesos contributed by the US Agency for International Development from its local currency PL 480 funds.[1] Administration of the operating fund rests with the Ministry of Agriculture and the Bolivian non-governmental institution.

One lesson from the Bolivian debt/nature exchange experience is that misperceptions may cause a nationalistic reaction. No foreign group acquired any control over land in Bolivia, although some press reports gave this impression. In fact, the land was already owned by the Bolivian government and has remained under government control. The debt/nature exchange is directed at supporting a more coherent long-term resource protection and management effort.

2. Costa Rica

Costa Rica created a debt exchange that permits the pooling of funds from debt donors in order to finance a multitude of programs. At first, the government set up a US $5.4 million debt/nature program in 1987 to establish a Natural Resources Conservation Fund using specially issued bonds that support the effort. The original $5.4 million ceiling was quickly surpassed by donor pledges, and, as indicated earlier, the total amount of conservation funds generated as of August 1990 was the equivalent of more than $40 million.

Many donors are involved. Funds to purchase debt have come from the Nature Conservancy, the World Wildlife Fund (including its national organizations from the US, Canada, and Europe), Asociacion Ecologica La Pacifica, the Pew Charitable Trust, the MacArthur Foundation, the JS Noyes Foundation, the Swedish Society for the Conservation of Nature, the W. Alton Jones Foundation, the Organization for Tropical Studies, and Conservation International.

One donor worth noting is the Fleet/Norstar Financial Group, which, in 1988 became the first financial institution to actually donate debt for a debt-for-nature exchange. Through the Nature

---

[1] "PL-480 funds" is a term for one of two types of debt owed to the US government. Funds can be authorized for international credit by the Foreign Assistance Act of 1961 (as amended), and funds are also authorized for international credit by Public Law 83-480, the "Agricultural Trade, Development and Assistance Act of 1954."

Conservancy, it donated more than US $250,000 (face amount) in Costa Rican debt paper towards conservation efforts in the Braulio Carillo National Park. This donation followed on the heels of a ruling by the US Internal Revenue Service that banks may donate debt for charitable purposes and deduct from their tax liability an amount equal to the market value of the local currency received in a conversion, with the balance of the face value allowable as a business loss.[2]

Costa Rican debt was selling for about 17 cents on the dollar in 1988, giving buyers a discount of more than 80 percent. In exchange for its notes, Costa Rica offered bonds at 75 percent of their face value. These bonds, in colones, with maturities up to six years, carry an average of 25 percent interest. Only the interest payments are immediately available to fund environmental projects because the bonds cannot be sold. Principal on the bonds isn't paid until the second year. When more cash is needed, however, the bonds can be used as collateral for loans. The swaps are administered by the Costa Rican Cooperative Bank (BANCOOP) on behalf of the National Parks Foundation (Fundacion de Parques Nacionales) and the Fundacion Neotropica, which channels the funds to projects agreed upon by the donors and the Ministry of Natural Resources, Energy, and Mines.

The newly generated conservation funds are being used to expand, manage, and protect many of Costa Rica's parks, building up the infrastructure for both tourism and science. The fixed costs of the parks, already set by law, take up to 90 percent of the existing government funds budgeted. No public money has been available for new programs, for reforestation, or for more environmental education work with the local population to prevent deforestation.

One major effort that is benefitting from the debt/nature exchange mechanism is the Guanacaste National Park Project. Completion of this park requires purchase of land in an area located between existing parks and reserves. When the park is completed, the Costa Rican government will assume title to it. The new park will be large enough to allow the restoration of an ecosystem that has all but disappeared elsewhere -- the dry tropical forest. Considerable technical assistance has come from the University of Pennsylvania.

Working with the local community is an essential part of Costa Rica's plan, so that people can perceive the park's value for them. Local non-governmental organization actively take part in the design and administration of programs funded by the debt/nature exchange.

3. Ecuador

Fundacion Natura, an important private conservation group in Ecuador, reached an agreement with the government's monetary board in 1987, enabling the Foundation to exchange up to US $10 million in debt for local currency bonds. These bonds were to be used to finance the Foundation's broad range of activities in conserving and improving Ecuador's national parks, among them Cayambe-Coca Ecological Reserve, Podocarpus National Park, Pasochoa Nature Reserve, and the Galapagos National Park.

---

[2]This ruling, formally known as Revenue Ruling 87-124, was issued in November, 1987 by the US Treasury Department. Its main purpose is to place the lender in the same position whether the debt is donated to charity or sold and then the proceeds are then donated to charity.

Fundacion Natura sought local and international donations to purchase this debt. The World Wildlife Fund-US arranged to purchase the first $1 million at just over 35 cents on the dollar; so US $354,000 bought $1,000,000 from commercial lenders to Ecuador.

The bonds were provided by the government at the full amount of the debt note, 100 percent of face value, and were converted to sucres at the official exchange rate. (The official exchange rate is substantially less than the floating rate.) The interest on these bonds, called Monetary Stabilization Bonds, is linked to the market rates -- currently about 33 percent annual interest -- and they mature over 9 years. The principal will become an endowment for Fundacion Natura, while the interest pays for its on-going programs.

Fundacion Natura is not the only beneficiary, because it will serve as a conduit for funds to support conservation efforts involving other non-governmental groups and public programs. The programs supported by these bonds will include management plans for protected areas, development of park infrastructure, acquisition of small nature reserves, and training of park personnel, in addition to broader environmental education activities.

Ecuador holds a wide variety of valuable habitats, from Amazon rain forests up to the Andean highlands and down to the coastal savannah and mangrove forests on the Pacific coast. Its Galapagos Islands, of course, are home to plants and animals found nowhere else on the planet. Until now, only the Galapagos Islands Park has produced net income. Tourism in the Galapagos alone has provided the budget for the other 12 Ecuadorian parks, which cover more than 10 percent of the country's land area. While the $10 million program may not substantially affect the country's debt, the first year's interest from the bonds has nearly doubled the national government's park budget.

Conclusions and Recommendations

The main attractive feature of the debt/nature exchange is that no money leaves the debtor country. In fact, there is not so much an "exchange" as a reallocation of government spending: from interest payments to foreign lenders to management of natural resources within the country. It is recommended that opportunities be explored for providing incentives to creditor nations' lending institutions to support this exchange mechanism. New tax incentives and revised accounting practices will help. The heart of the challenge, it seems to me, is to provide lending institutions with assurances that the funds will be used properly to achieve stated conservation goals. This is as much a problem in cross-cultural perception and communication as it is a technical problem. One major contribution to overcoming this problem may be to enhance the capacity of local institutions to control conservation initiatives.

Participation by non-governmental organizations is essential, and governments must actively seek to provide incentives for such participation. Opportunities exist for direct federal intervention, like proposed Congressional legislation and the Enterprise for the Americas Initiative that US President

Bush proposed in June, 1990.[3] Governments can also play an indirect role. Promotion of technology transfer is recommended as one such way to encourage NGO participation. In particular, a technology transfer can result in better data collection and management for habitat value and land tenure, and a better understanding of their interaction. In this way, proposed conservation programs can be more firmly rooted in current knowledge of the programs' social, economic, and biophysical setting, and the organizations that generate such knowledge will be empowered by the very act of doing so.

The Initiative aims to promote economic gains for Latin America and Caribbean countries and the US. Eligibility for any form of debt forgiveness and restructuring depends on meeting certain economic criteria. These criteria for eligibility include: (1) an IMF/World Bank economic reform program; (2) various trade and investment reforms; and (3) satisfactory agreements with commercial banks.

If these conditions have been met, then a country would be eligible for partial debt forgiveness and restructuring, and then finally, the establishment of environmental trusts. Because of these stringent conditions, however, many countries (e.g., Brazil and Peru) would not yet be eligible. For an eligible country, the US government would forgive part of the debt and permit the interest on the remainder to be paid into a local currency trust fund at a concessionary interest rate. Then, monies would be disbursed from the trust for environmental programs.

---

[3]The Enterprise of the Americas Initiative was first announced by the Bush Administration in June, 1990. There are three main parts to the Initiative: (1) Trade -- establish free and fair trade rules such as reciprocal reductions in trade barriers; (2) Investment -- introduce major investment reforms that might include privitization of state-owned companies and the liberalization of investment regimes. The Initiative calls for the Inter-American Development Bank to play a significant role in effecting these reforms; and (3) Debt -- permit partial debt forgiveness of sovereign debt owed to the US government for eligible countries in Latin America and the Caribbean. This does not apply to debt owed to commercial banks.

Participation of the Communities in the Protection of the Natural Habitat of the Neotropical Primates and Natural Resource Policy in Bolivia

Hernán Zeballos H.

Undersecretary for Agricultural Production
Ministry of Rural and Agricultural Affairs (MACA)
LaPaz, Bolivia

1. IMPORTANCE OF RENEWABLE NATURAL RESOURCES

i. In the conservation of ecosystems.
ii. In the proper management of resources.
iii. In the specific benefits of each individual resource.
iv. In the integration of man as a natural resource.

2. IMPORTANCE OF THE NONHUMAN PRIMATES

With respect to ecology:

i. For the dispersion of fruits and propagation of seeds.
ii. For the relationship between flowering and fruiting.
iii. To assist in defining the presence of vegetative associations.
iv. For the influence and interrelationship between certain animal species and their ecological position.
v. To influence the dynamics of the ecosystem in the forest.

With respect to science:

i. Because man is a primate there is a strict phyllogenetic relationship between him and the nonhuman primates.
ii. Because the nonhuman primates can serve as execellent models for biomedical research.
iii. To aid man to understand his anatomy and physiology and even his social behavior.

With respect to culture:

i. Because the nonhuman primates are so similar to man they constitute didactic models for the teaching of the biological sciences.
ii. For their use in play as mascots and as zoological species in which man finds complementarity in his understanding of a life similar to his.

With respect to social matters:

i. For their epidemiological role as vectors of diseases communicable to man, such as malaria, yellow fever, and others.
ii. For their economic importance for the inhabitants of the primary selva, as a part of their sustenance, which should be regulated at a later stage as a possible source of income if the animals are managed properly in zoological nurseries or nursery reserves.

3. ROLE OF THE PRESERVATION OF SPECIES FOR THE WELL-BEING OF HUMANITY

i. First, they have an ethical role through the relationship of man and the environment, since man himself is the fruit of biological evolution.
ii. To increase the relatively limited understanding by man of the biological diversity of the planet and as a focus to preserve the maximum biodiversity that could bring man unsuspected benefits in the near future.
iii. To maintain the ecological equilibrium, since they play specific roles; for example, the minuscule bacteria and the large species, such as whales and elephants, exercise an influence on the rest of the animal and vegetable species.
iv. Constituting, in part, the natural heritage of the countries and an intangible richness, but possibly in the future, if our knowledge of it increases, a real economic resource.
v. To manifest the ethical values of present generations, the importance of leaving to future generations a legacy similar to the one they received.

4. IMPORTANCE OF THE PRESERVATION OF THE NONHUMAN PRIMATES IN THE BOLIVIAN CASE

In the universal context indicated above it becomes necessary to preserve that half-brother of man, the nonhuman primates, with greater reason in Bolivia for:

i. Ecological diversity

Bolivia, utilizing the Holdridge classification zones of life, has:
3 principal physiographic regions and
27 zones of life.

ii. Having 557,000 km^2 of forests (somewhat more than 50% of the national territory) in 2 large regions

iii. The potential richness in wild flora and fauna that the extensive regions of Amazon and non-Amazon forests represent.

In terms of fauna, it is known that there are approximately
1,270 species of birds,
300 species of mammals,

more than 800 species of fish, and
approximately 100 species of amphibians.

In terms of flora, it is estimated that there are more than 20,000 to 22,000 species[1].

In this magnificent diversity it is estimated that there are approximately
18 species of nonhuman primates and
22 subspecies[2].

iv. The irrational use that is made of the abundant, diverse primates in Bolivia and because these are insufficiently known.

The irrational use is manifested in activities whose purpose is feeding as well as the hunt to obtain mascots; and for activities oriented toward the exportation of animals to the developed countries.

The primates are also affected by the processes of modification of habitats, modifications that reach grand proportions and are due principally to the lack of planning in the development and colonization of the wilderness areas.

v. The scarce knowledge available on these species as noted by Heltne.[3] Field studies on nonhuman primate populations in Bolivia are rare and remote in time.

The first studies were carried out in La Paz by a German museum which sent a mission to collect animals. Fifty-five years later a mission sponsored by PAHO/WHO[4] carried out research in the Departments of Santa Cruz, Beni, and Pando, and in the northern part of La Paz.

More recently Isawa (1979-1981), Yonida (1981-1984), Masataka (1981), Minisawa (1981-1984), Poola (1979-1982), and Garcia (1986-1988) carried out field work--the first four in the Departments of Beni and Santa Cruz.

---

[1]Beck, Stefan, Director Herbario Nacional, personal communication.

[2]Anderson (1985). Lista preliminar de mamíferos bolivianos. Cuadernos Academia Nacional de Ciencias de Bolivia. VI Ciencias de la Naturaleza Zoología 3, La Paz. 16 pp.

[3]Heltne, Paul (1975). "Investigación en el terreno sobre poblaciones de Primates no Humanos en Bolivia." OPS/OMS. 25 pp.

[4]Anderson (1985). *op. cit.*

5. BOLIVIAN POLICY WITH RESPECT TO NATURAL RESOURCES AND THE ENVIRONMENT

Because of its location in the center of the South American continent Bolivia shares its different ecosystems. It is a country of Andean tradition and image. In reality the greatest part of its territory consists of tropical lowlands, belonging to the Amazon and La Plata River basins. To its geographical diversity is added its cultural diversity. A complex historical process has maintained the presence of numerous native villages in the Andean as well as the Amazon regions of the country.

The growing deterioration of the ecosystems provoked by the application of inadequate models of development, in which the sustained use of the resources was not taken into account, led to the current deterioration of the environment and the irrational exploitation of the renewable natural resources. In the face of this situation the Government of Bolivia is executing a development policy in which the environmental factors of preservation as well as sustained use are incorporated in the economic and social development of the country.

Observance of a "Historic Ecologic Pause of Five Years" has been decreed; it prohibits new concessions for forest products, hunting, and pursuit of wild animals and regulates actions for better control and conservation of the environment.

i. Ecological Pause

It is based on the need to establish a period of time essential for allowing reordering of all those processes that, because of their impact on the natural world, endanger the sustainability of the material base of human life and the natural Bolivian heritage.

Furthermore, we believe that the natural resources of our country can no longer have the inevitable destiny of their destruction as the sole road to the elevation of the quality of life of the Bolivian people. Our environment constitutes a productive potential to which we assign the greatest importance, for the collective benefit of our country and the world.

The historical ecological pause marks a milestone in the dynamics of the processes of development and their relationship with conservation in Bolivia.

This constitutes a pledge and an unknown challenge to the Bolivian people and their history. But it also assumes an awakening in Bolivia to the concern and actions that the international community manifests on this subject: a commitment contracted with it.

We accept the criterion of international coresponsibility in the face of global deterioration of the environment. World development models show maladjustments. Progress and technology have also brought environmental degradation and, what is worse, have not resolved the problem of marginality and poverty in countries such as ours. But we also understand that, on the basis of shared responsibility, international relationships can be established which are more creative and more just and which allow us to face the future with hope.

ii. Administration of Natural Resources

The administration of the natural resources and the environment is effected through the Undersecretary of Renewable Natural Resources and the Environment, whose objective is the standardization and supervision of national policies and strategies on the environment and natural resources through planning and programming of activities for execution, control, and evaluation of water, forest, soil, air, plant, animal, and hydrobiological resources. Similarly, its function includes the application of technical operational standards for actions for control of the environment and protected areas in rural and urban areas.

Units that constitute the Office of the Undersecretary:
- National Administration of the Environment
- National Administration of Integrated Resources
- National Administration of Fishery Resources
- Forestry Development Center
- Fishery Development Center
- Bolivian Indigenist Institute

Fields of activity:

The fields of activity of the Subsecretariat are those of research, planning, control, standards, and supervision of the natural resources and the environment, including actions for control and protection of the forests, protection of the wild fauna, conservation of national parks, control of environmental contamination, preservation of the soil, water, animal, plant, and atmospheric resources.

Thus it formulates and directs the policies for the indigenous and tribal populations of the country.

iii. Law of the Environment and Natural Resources

The adoption of the Law of the Environment and Natural Resources constitutes one of the principal objectives of the government, since its contents will regulate all the factors involved in the environmental problem (flora, fauna, air, water, and soil) in an integrated manner.

It will be a modern instrument that will allow the restructuring of the entire state apparatus related to the environment in order to achieve the sustained benefits of the resources and to preserve the natural environment for future generations.

The Law of the Environment is currently being reviewed by a special high-level commission and it is expected that it will be promulgated within eighteen months.

6. BOLIVIAN POLICY ON RENEWABLE NATURAL RESOURCES

i. Administration of Renewable Natural Resources

The Center for Forest Development is the organism that determines policy on the subject.

Its functions cover: the protection, management, utilization, transport, and marketing of wild animals and the products that can be obtained from them as well as the protection of the species threatened with extinction, the conservation of the habitat for the fauna and flora, and the establishment of national parks and other activities aimed at the conservation, promotion, and rational exploitation of the natural resources.

ii. The project of the Law of the Environment is in the process of being approved by the National Congress.

iii. Status of protection of primates.

These are protected indirectly by two international provisions, the Convention on International Trade in Endangered Species (CITES) and the Red Data Book with the species protected by the IUCN.

Thus it is of great interest for the Government of Bolivia and the Ministry of Rural and Agricultural Affairs (MACA) to establish specific programs and projects, such as the Bolivian national agreement on primatology to be signed with PAHO/WHO which would make it possible to improve the legal framework and the conservation and economic production of natural resources.

7. PARTICIPATION OF THE INDIGENOUS COMMUNITIES IN THE CONSERVATION OF THE NATURAL HABITAT OF THE NEOTROPICAL PRIMATES

Bolivia has recently been troubled by the indigenous march "for dignity and territory."

Approximately 800 indigenous persons carried out a march of 620 km from the cathedral in Trinidad, Beni, to the cathedral in the city of La Paz.

The execution of this march, which expresses the pressures to social groups of greater economic power (cattlemen and forestry enterprises), and the ethnic groups and migratory movements of the inhabitants of the rural areas from the high Andean zones toward the subtropics required the central government to make legal provisions in favor of the ethnic groups from the eastern part of Bolivia.

Four Supreme Decrees, Nos. 22609 to 22612, declare the national, social necessity for recognition, assignment, and occupation of territories in favor of groups from the selva and indigenous communities from the eastern and Amazon regions.

These decrees also recognize the international agreements that consider the rights of the indigenous populations to the natural resources existing in their lands and to special protection.

It is in this new legal framework that at least two interesting possibilities appear: the first, to benefit from the ability of the groups in the selva to coexist in the forests and with the forests, and the second, to establish projects that, without disturbing the equilibrium of the ecosystems, exploit the resources of flora and fauna for regular commerce, particularly scientific, thus generating income for those communities.

The projects linked to these ethnic groups can thus serve the following proposals:

i. The generation of income.

ii. To serve as a link between the communities and the state powers to provide health and educational services.

iii. To provide technical assistance to new forms of agricultural, forest, and pastoral production.

iv. To return to the communities the results of the medical, ecological, and other research related to the projects.

Regional Plan of Action in Primatology for the Americas - A Proposal

Filomeno Encarnación

Hugo Samamé

Center for Reproduction and Conservation of Non-human Primates
Preuvian Primatological Project
Veterinary Institute of Tropical and High Altitude Research
IVITA
Universidad Nacional Mayor de San Marcos
P.O. Box 621
Iquitos, Peru

I. INTRODUCTION

The 157 Member States of the World Health Organization collectively expressed their concern, during the Twentieth Eighth and Twentieth Ninth World Health Assemblies, for the danger encountered by species of nonhuman primates because of the destruction of their natural habitats. In Resolutions WHA 28.83 and WHA 29.67 the WHO Director-General is directed to facilitate the exchange of resources and technology among the affected countries and, with the assistance of international agencies, provide advisory services on the conservation, rearing, and reproduction of the species.

The official representatives of Peru, Mexico, Canada, the United States of America, Brazil, Guatemala, Spain, and South Africa, staff members of the United States Agency for International Development, investigators in universities in the United States and Peru, and representatives of nongovernmental organizations in Peru that attended the "Workshop on Controlled Breeding of Nonhuman Primates in Their Own Habitat" held in Iquitos in November 1986 signed the "Declaration on Technical Cooperation among the Countries of the Americas for the Conservation and Utilization of Nonhuman Primates." In that document it is recommended that a Regional Committee on Primatology for the Americas be created on the basis of the experience developed by the Peruvian Primatology Project "Manuel Moro Sommo;" at the same time the collaboration of PAHO is requested for the preparation of the regional strategies and plans of action promoting intercountry cooperation that make possible the development and strengthening of the national programs for the conservation, reproduction, and utilization of the neotropical primates.

In RIMSA V, held from 27 to 30 April 1987, the Peruvian delegation requested consideration of the "Declaration on Technical Cooperation among the Countries of the Americas in the Conservation and Utilization of Nonhuman Primates;" thus in Resolution VIII it is recommended that PAHO, acting as ex-oficio secretariat, coordinate the establishment of the Regional Committee on Primatology.

IMPORTANCE OF THE REGIONAL PLAN

The Regional Plan constitutes a guideline and frame of reference within the political will of the countries of the Region which frames the national plans for the conservation and protection of the species of neotropical primates as well as their natural habitat, seeking concurrence for the regional effort and facilitating the definition of actions and common causes among the Member Countries.

The Regional Plan provides the fundamental orientation for the policies on conservation, organization of procedures, and cooperative intercountry programs among Members of PAHO. In the world context, it represents the participation of the region in the implementation of global strategies of conservation and utilization of the nonhuman primates in their own habitat, compatible with Resolutions WHA 28.83 and WHA 29.63 of the WHO Assemblies.

Considering a certain homogeneity and similarity in the ecology and dynamics of the system of humid tropical forests in the neotropics, the goals and the objectives of the Member States for conservation are consonant with their particular situations and possibilities, so that through joint effort the conservation and perpetuation of the nonhuman primates can be guaranteed.

II. SITUATION OF THE PRIMATES IN THE NEOTROPICS

The life of the natural populations of neotropical primates is directly related to the presence of the natural forests, in whose vertical and horizontal spaces these animals carry out their functions of reproduction, feeding, and resting, among others. However, the presence of the forest in the humid tropics in general depends on the degree and level of understanding of the human population for its conservation and protection.

Large extensions of the forests are being modified by the use of the soil in orthodox and traditional ways, in some cases so accelerated through the use of machinery and equipment that man himself is inventing for application in agriculture or tree-felling from the temperate areas that the soil is disappearing or being degraded.

We should be aware that the tropical humid forest constitutes the richest ecosystem of the land and regret that it covers only 8% of the total area of our planet. In this forest live between 40 and 50% of the species whose total number is estimated at 5 to 10 million. But, this biome is less known than the others, except for the nonhuman primates, which are indeed rather well-known, although the loss of the forest habitat is a predominant factor in the reduction in the numbers of the those species.

Humankind experiences moments of strong, algid discrepancies between two concepts of the use of the forest. On the one hand it is affirmed that the greatest disturbance or alteration leads to impoverishment, and on the other, that clear-cutting favors its recovery and optimizes its utilization. The following estimates are given by two investigators for the 1970s: 200,000 km^2/year of tropical forest were affected by degradation and destruction through anthropic action (Myers, 1980) and 76,000 km^2 of forest were eliminated each year as a result of the sale of wood (Lanly, 1982).

In Latin America the principal agents of deforestation are the upswing in livestock raising which causes the clearing of some 20,000 km^2/year and the monocultural plantings of oil palms, tea, cacao, coca, sugarcane, plantains, and others, which in a great percentage of the cases continue to produce for up to 50 years.

In future decades the loss or impoverishment will markedly affect the continuation of valuable critical work on evolution. We warn the environmental services of the loss of the layer of nutrients of the soil, disorganization of the hydrologic systems and water sources, and other dangers that are reflected in the climatic changes through the alterations in the CO_2 cycle in the global atmosphere so that the life of the primates is affected. The examples in the southern and eastern parts of Brazil (Rondonia, Pará, Matto Grosso, and Acre) are due to colonization with livestock; on the Atlantic coast of Brazil, to the removal of wood and to agriculture; in the Brazilian stretch of the trans-Amazon road; in the stretches of marginal road in the forest, Alto Mayo, and Huallaga Central of Peru; in Central America, to agriculture, livestock-raising, and removal of wood; in the Amazon basin (Caquetá and Putumayo) of Colombia, to colonization and livestock raising; on the Pacific coast and in the Amazon region of Ecuador, to agriculture and removal of wood; and in the Amazon region (upper and lower selva) of Peru, to cultivation of coca and livestock-raising.

In the forests with limited accessibility that constitute habitats for nonhuman primates, evaluations of their populations have still not been initiated. But the felling has had a drastic effect in some countries, while the methods and procedures for analysis and sampling of the populations of primates are still being perfected.

The natural populations of primates in the forest with its patterns of geographical distribution, abundance, and vulnerability are related to aspects of the biology of the species. Thus body size is directly related to the period of gestation and the size of the territory, but indirectly to population density. In other words large size presents greater options for predation by man and also offers more food.

Technological progress, with machinery and equipment in accordance with the scale, level, and nature of its application, can help us to understand the implications for the exploitation or utilization of the forest. Destructive exploitation of species of plants, such as *Castilloa ulei* (for latex) and *Couma macrocarpa* and *Ragala sanguinolenta* (for rubber) in the Amazon basin, affects the dietary resource of *Chiropotes* and *Lagothrix*. The traditional collection (felling) of Amazonian native fruit trees, such as *Rheedia*, *Spondias*, *Inga*, *Caryocar*, and *Parahancornia*, decreases the possibilities for food for *Saimiri*, *Saguinus*, *Cebus*, and numerous birds and mammals. Clearing the underbrush for exploitation at the base of the trees of *Hevea brasiliensis* (rubber plant), and of the liana *Paullinia cupana* (supplejack) or felling for the preparation of firewood or charcoal in the proximity of the hamlets reduces the entomologic fauna for the diet of *Saimiri*, *Cebus*, *Saguinus*, *Callithrix*, *Callicebus*, and *Callimico* in the Amazon region.

The construction of roads and dikes without adequate technical structures and recommendations causes impounding, yielding floods that kill the vegetation or forest. Examples are the hydroelectric dams at Tucurui on the Tocantins River and Las Balbinas in Roraima, Brazil, where there was flooding and the death of extensive forests inhabited by primates and other tree-dwelling animals.

Agricultural use of the forest progresses from the mountainous forests toward the plains, so there is an increase in the proportion of secondary forest that will induce changes in the structure and population dynamics (community) of primates. In Colombia there are populations of *Lagothrix lagothricha* and *Ateles belzebuth* in isolated forests and *Cebus albifrons* in secondary forests where *Saguinus leucopus* also prosper. In Manaus, Brazil, in 20 Ha. of isolated forest, *Pithecia* and *Saguinus bicolor* survive, the latter under normal conditions in the secondary forest. Incidentally, these animals acquire semiterrestrial behavior to cross the roads and cleared areas.

The selective felling of commercial wood also affects the forest through the procedures and traditional mechanisms applied, such as concessions in areas without knowledge of the natural components, the felling and system of transportation by devastating dragging toward water masses, and the opening of forest roads and trails passable by vehicular traffic. These are common practices in the Amazon region for extraction of mahogany (*Swietenia macrophylla*), cedar (*Cedrela* spp.), "ishpingo" (*Amburana cearensis*), "cumalas" (Family Myristicacea), and others. These trees are important as substrata; however, the exudates and rubber from cedar (*Cedrela* sp.) and "quillusisa" (*Vochysia* sp.) in Peru are principal components of the diet of *Cebuella pygmaea* and the fruit of *Virola* (Myristicacea) is important for the feeding of *Cebus apella* and *Alouatta*.

In addition, the state and private livestock and agroindustrial enterprises, called poles or foci of development or occupation of the Amazon region, promote the clearing of large areas with total elimination of the arboreal stratum and proliferation of the herbaceous stratum because of the nature of the crops chosen--for example, pastures based on legumes and grasses, rice in unirrigated arable land or in puddles, peanuts, corn, cotton, tobacco, pepper, and others. The consequences are drastic for the tree-dwelling animals because of the disappearance of hiding-places and refuges, the substratum, and the sources of food. The effect is maximized because the seeds and all possibility of regeneration are eliminated. These examples are visible in different parts of the Amazon regions of Brazil, Peru, and Colombia.

The hunting of nonhuman primates in the neotropics affects mainly the large species, such as *Ateles*, *Lagothrix*, *Alouatta*, and *Cacajao*, which are sources of animal protein for the native inhabitants or colonists. Between Peru and Colombia it is common practice to use *Lagothrix* as bait to hunt felines (*Felis pardalis*) the skins of which are valuable. The ornamental use of skins, feet, and skulls is common in the Amazon region for preparation of wall decorations, keyrings, necklaces, etc. from small species, mainly *Saguinus*, *Saimiri*, and *Callicebus*. The tails of *Pithecia* are popular as plumes.

The hunting of parents or adults in order to obtain animal mascots leads to a great deal of interest in the *Saguinus*, *Cebuella*, *Callithrix*, *Leontopithecus*, *Callimico*, and *Callicebus*, while other species, such as *Saimiri* and *Cebus*, are hunted as a sanitary measure since they invade fields of corn, cowpeas, plantains, and various fruit trees.

Capture for the mascot trade or, in this century, for use in laboratories continues to play a preponderant role and has an impact on the primate populations. However, in July 1975 the governments of 80 countries, members of the IUCN, established guidelines and agreements to control this trade while at the same time protecting the species in the process of extinction with the signing of the Convention on International Trade in Endangered Species (CITES) in Washington. In the neotropics, commercial demand, mainly from the Amazon region, had its beginnings in the 1940s, with the United States of America as the first country to receive or import. To cite some figures as examples, from Peru over a period of 11 years (1964 to 1974) almost 340,000 live individuals were imported, a figure that has decreased notably from 1976 to 1989 through control measures of the Government of Peru as it carried out the CITES recommendations and eliminated causes of mortality occurring from the time of capture until arrival at the final destination. These results are due to the active participation, coordination, and advisory services of PAHO/WHO.

The neotropical primates inhabit territories of the countries of Central and South America, except Uruguay and Chile, and some of the Caribbean islands, from southern Mexico through Central America to the southern extreme of Brazil and northern Argentina,. The distribution of the species corresponds to the concept of biogeographical regions for the neotropics.

According to Mittermeier (1987), there are 16 known genera with some 65 species; the latter were recently increased with the studies published by P. Hershkovitz of the Museum of Natural History in Chicago, M. Ayres of the Emilio Goeldi Museum in Belém, and others. Two genera, more than 12 species, and a total of 30 subspecies are considered to be threatened with some of them actually on the way to extinction. (See tables.)

The scientists and authorities of countries with territories that maintain natural populations of primates carry out efforts that are national or internationally coordinated in order to gain financial assistance that involves protection of the species through the systems of conservation units.

The greatest specific diversity is in the Amazon region, with 12 genera and more than 40 species, and in the Atlantic forest of Brazil, with 6 genera and 15 species occupying preferential habitats below 1,000 m above mean sea level. Only *Lagothrix flavicauda*, a species endemic to Peru, inhabits territory above 1,000 m above mean sea level, on the line that divides the Andean from the Amazon region. In the endemism of the species resides the importance of the Amazon forests and the Atlantic coast of Brazil, where national institutions, with contributions from foreign and international organizations, are making efforts to evaluate the populations and establish conservation units.

Countries, such as Brazil with more than 51 species of primates and some 35 endemic species, Peru with more than 28 species and 11 endemic species, and Colombia with more than 27 species and 7 endemic species, have assumed leadership for the protection of their primatologic fauna and the conservation of their habitats in national parks, community reserves, areas under study, and elsewhere.

On the Atlantic coast of Brazil, since 1979 priority has been given to work to safeguard the genera *Brachyteles* and *Leontopithecus*, considered threatened by the destruction of the habitat. The results in captivity and in the natural habitat are contributing to the delineation of certain patterns for the recovery of the forests and the protection of the remaining forest, which also houses *Cebus apella xanthosternos*, *Callicebus personatus melanochis*, *Callithrix kuhli*, and *Alouatta fusca*. The actions were taken by the Brazilian Foundation for the Conservation of Nature (FBCN), the Federal University of Minas Geraises, and the Primate Center in Rio de Janeiro, mainly with the support of the WWF.

III. GENERAL OBJECTIVE

To safeguard the continuing existence of the wild populations of nonhuman primates in their natural habitat so that the benefits are shared and there are improvements in aspects of human living conditions, such as health, education, and agriculture.

IV. STRATEGIES AND ACTIONS

1. DECLARATION OF NATURAL AREAS FOR THE PROTECTION OF THE WILD FAUNA AND ESPECIALLY FOR THE CONSERVATION OF NONHUMAN PRIMATES.

   1.1 To carry out inventories and evaluate the ecosystems with native populations.

   1.2 To prepare maps of national and regional geographical distributions.

   1.3 To select and establish protected areas so that all the species are safeguarded.

2. PROMOTION AND DEVELOPMENT OF INTERSECTORAL COORDINATION FOR THE JOINT EXECUTION OF PROGRAMS FOR THE EVALUATION OF HABITAT, FOR REPRODUCTION, AND FOR CONTINUING SURVEILLANCE OF RARE SPECIES OR THOSE BECOMING EXTINCT.

   2.1 To prepare and carry out national plans of action with the technical cooperation of PAHO when it is required.

   2.2 To develop and expand the concept of a policy on conservation and management, including the maintenance of specific populations and their variations that have current and potential socioeconomic significance.

   2.3 To develop and adopt methods and techniques of management and sustained utilization of nonhuman primates as a natural resource produced by the forest.

   2.4 To prepare standard formats to facilitate technical and scientific understanding between specialists and administrators and the financial agencies and environmental organizations.

3. COOPERATION BETWEEN COUNTRIES FOR THE CREATION OF JOINT AREAS FOR THE PROTECTION OF SPECIES OF NONHUMAN PRIMATES AND OTHER WILD ANIMALS.

   3.1 To sign agreements, and/or bilateral treaties between neighboring countries.

   3.2 To formulate PAHO technical cooperation in the execution of the intercountry primatology programs.

   3.3 To establish agreements between countries for the transfer of technology.

4. DISSEMINATION OF TECHNICAL AND SCIENTIFIC INFORMATION TO STRENGTHEN THE DEVELOPMENT OF PERSONNEL AND THE INFORMATION SYSTEMS IN DISCIPLINES RELATED TO PRIMATOLOGY.

4.1 To establish and identify documentation and/or reference centers for neotropical primatology.

4.2 To train specialists at the academic and technical levels.

4.3 To establish basic primatological information systems for the conservation of genes and their natural habitat.

4.5 To promote the importance of the conservation of the ecosystem and its genetic resources at all levels.

4.6 To publish the results of studies and unpublished documents, facilitating dissemination and exchange.

5. NATIONAL AND INTERNATIONAL INTERSECTORAL COOPERATION FOR THE FORMULATION AND APPLICATION OF ENVIRONMENTAL EDUCATION PROGRAMS.

5.1 To incorporate in the programs and curricula at all levels material related the biology of primates.

5.2 To promote the participation of the human population in urban and rural areas in the conservation and protection of the various species and the ecosystem.

6. ESTABLISHMENT OF CENTERS FOR REPRODUCTION OF NONHUMAN PRIMATES IN CAPTIVITY AND/OR SEMICAPTIVITY FOR PURPOSES OF REPOPULATION AND SUSTAINABLE USE IN BIOMEDICAL RESEARCH.

6.1 To develop standard plans for research, management, and reproduction in captivity.

6.2 To develop policies for humanitarian management, quarantine, conditioning, and production of nonhuman primates for scientific use in their native countries and for transfer to third countries.

6.3 To evaluate the potential risks of zoonoses in neotropical primates.

6.4 To establish genetic lines of primates.

6.5 To strengthen trials of the adaptation of species bred or maintained in captivity to natural habitat.

7. MOBILIZATION OF NATIONAL AND INTERNATIONAL ECONOMIC RESOURCES FOR THE DEVELOPMENT OF NATURAL AREAS, CENTERS FOR REPRODUCTION OF NONHUMAN PRIMATES, AND PLANS AND PROGRAMS FOR CONSERVATION.

7.1 To obtain financial resources for training in primatology and implementation of urgent, critical, or long-term studies and/or research.

7.2 To formulate national and multinational exploratory projects jointly with the bilateral or multilateral financial agencies.

7.3 To establish a regional fund for the implementation of actions related to the conservation of neotropical primates, as well as national funds for the same purpose.

Section 2

SPANISH

Sección 2

ESPAÑOL

Memorias de la Primera Reunión
del Comité Regional de Primatología
para las Américas (CORP I)

PRIMATES DE LAS AMERICAS

ESTRATEGIAS PARA LA CONSERVACION Y USO SOSTENIDO EN INVESTIGACION BIOMEDICA

EDITADO POR

Primo Arámbulo III
Filomeno Encarnación
Jaime Estupiñán
Hugo Samamé
Charles R. Watson
Richard E. Weller

Centro de Conferencias de Battelle
Seattle, Washington, EUA

29–31 de octubre de 1990

PATROCINADO POR

Organizacion Panamericana de la Salud
Organizacion Mundial de la Salud

BATTELLE PRESS
Columbus • Richland

INFORME FINAL

I Reunión Ordinaria del Comité Regional de Primatología para las Américas

(CORP 1)

Seattle, Washington, EUA

29-31 de octubre de 1990

INFORME FINAL

La I Reunión del Comité Regional de Primatología para las Américas se celebró en la sede del Battelle, Centro de Conferencias de Seattle, Estado de Washington, Estados Unidos de América, de acuerdo a la convocatoria hecha por el Director de la Oficina Sanitaria Panamericana, en cumplimiento de la Resoluciones VIII y XIV de la V y VI Reuniones Interamericanas de Salud Animal a Nivel Ministerial, respectivamente y posteriormente ratificadas por los Cuerpos Directivos de la Organización Panamericana de la Salud (OPS/OMS).

JUNTA DIRECTIVA

Por unanimidad, la Junta Directiva de la Reunión quedo constituida así:

Presidente:	Ing. Manuel Uceda Director General de Forestal y Fauna Ministerio de Agricultura Perú
Vicepresidente:	Dr. Stephen Potkay Director Veterinary Resources Program National Institutes of Health United States of America
Vocales:	Dr. Alejandro Estrada Estación de Biología Tropical Universidad Nacional Autónoma de México México
	Dr. Jordán Paulo Wallauer Jefe División de la Fauna y Flora Silvestres Secretaría Nacional de Medio Ambiente Brasil

El Dr. Primo Arámbulo III, Coordinador del Programa de Salud Pública Veterinaria OPS/OMS, en representación del Director de la Oficina Sanitaria Panamericana, Dr. Carlyle Guerra de Macedo, actuó como Secretario *ex officio* y Relator.

PARTICIPANTES

Los siguientes Gobiernos, miembros natos del Comité, estuvieron representados en la Reunión: Argentina, Barbados, Bolivia, Brasil, Canadá, Colombia, Estados Unidos de América, México, Paraguay y Perú.

Asistieron como observadores representantes de instituciones de los siguientes países: Brasil, Colombia, Estados Unidos de América, Francia, Guatemala, Japón y Países Bajos, así como el "International Council for Laboratory Animal Science".

La firma Battelle Pacific Northwest Laboratories actuó como institución anfitriona.

La lista de participantes se adjunta como Anexo 1.

PROGRAMA DE TEMAS Y PROGRAMA DE SESIONES

La Reunión celebró una sesión inaugural, tres sesiones plenarias y una sesión de clausura.

La primera sesión plenaria restringida para los miembros natos de la CORP eligió la Junta Directiva y nombró una comisión integrada por los delegados de Argentina, Colombia y Paraguay para que estudiaran los proyectos de Estatutos y Reglamentos de la CORP.

En esta misma sesión se constituyeron otras tres comisiones. La primera sobre movilización de recursos, bajo la presidencia del Dr. James Wong de Canadá; la segunda sobre investigaciones en primatología, bajo la presidencia del Dr. Julio C. Ruiz, de Argentina, y la tercera sobre conservación de primates no humanos, bajo la presidencia del Dr. José Vicente Rodríguez, de Colombia.

Luego se aprobó el programa de sesiones. (Ver Anexo 2).

En la sesión inaugural el Presidente cedió la palabra al Dr. Roger Anderson de Battelle Pacific Northwest Laboratories, quien en nombre de esa institución manifestó la satisfacción de colaborar en la realización de la Primera Reunión del CORP y asimismo informó que Battelle es consciente del rol importante que desempeñan los primates en la ciencia.

El Dr. Primo Arámbulo III en representación del Director de la Organización Panamericana de la Salud, dió un saludo muy especial a las delegaciones y observadores participantes a la reunión e informó los objetivos de la convocatoria y puso en relieve la política de OPS/OMS en relación a la conservación y uso sostenible de los primates no humanos y su importancia en la salud pública.

El Presidente solicitó al Dr. Jordán Wallauer hacer uso de la palabra, quien en nombre del Secretario Especial del Medio Ambiente de Brasil manifestó la complacencia por la realización de esta iniciativa. Posteriormente, el Presidente pidió al Dr. Hernán Zeballos, Subsecretario de Desarrollo Agropecuario de Bolivia, inaugurar la Reunión. El Sr. Subsecretario manifestó que la participación a esta I reunión del Comité permitirá encontrar alternativas de control a la destrucción del medio ambiente, la que se asocia a la pobreza de las poblaciones situadas en las regiones selvícolas, e hizo votos para que la I Reunión del CORP alcance los objetivos propuestos.

En la segunda sesión plenaria sobre conservación, manejo y salud se trataron los temas "Estado Actual y Conservación de los Primates Neotropicales en su Propio Habitat" presentado por el Dr. Roderic Mast de "Conservation International"; "El Estado de las Poblaciones Naturales de Primates y los Beneficios de la Cosecha Sostenida" por el Biólogo Filomeno Encarnación, Asesor Temporal de OPS/OMS; "Los Factores Ambientales para la Reproducción de Primates en Habitat Especiales" por el Dr. Thomas Wolfle, de la Academy of Science, de los Estados Unidos. "La Conservación de los Primates No Humanos y su Importancia en la Salud Pública", por el Dr. Stephen Potkay, Jefe Servicio de Recursos Veterinarios de los Institutos Nacionales de Salud de Estados Unidos de América; y "El Programa de Cooperación Técnica de la Organización Panamericana de la Salud", por el Dr. Primo Arámbulo III, Coordinador del Programa de Salud Pública Veterinaria de la Organización Panamericana de la Salud (OPS/OMS).

El Presidente cedió la palabra a los delegados de Argentina Barbados y Paraguay para que informaran sobre la primatología en sus respectivos países.

En la tercera sesión plenaria se trataron los temas relacionados a política y economía. El tema "Manejo de Fauna en Brasil" por el Dr. Jordán Paulo Wallauer, de la Secretaría Especial de Medio Ambiente, Brasil; "Manejo de los Primates No Humanos como Recurso Renovable: La Economía de la Conservación" por el Dr. Primo Arámbulo III de la OPS/OMS; "El Desarrollo Integrado de las Unidades de Conservación y el Rol de los Programas Nacionales de Primatologia" por el Dr. José Vicente Rodríguez, del Instituto de Recursos Naturales, de Colombia; "La Cooperación Multisectorial e Interpaises para el desarrollo de Programas Nacionales de Primatologia" por el Ing. Manuel Uceda, Director General de Forestal y Fauna del Ministerio de Agricultura del Perú; y "La Movilización de Recursos para Conservación del Ecosistema" por el Dr. Edward Liebow, de

Battelle Pacific Northwest Laboratories; "La Participación de la Comunidades en la Protección del Habitat Natural de los Primates Neotropicales y la Política de Recursos Naturales de Bolivia" por el Dr. Hernán Zeballos, Subsecretario de Asuntos Agropecuarios del Ministerio de Asuntos Campesinos y Agropecuarios de Bolivia; "El Plan de Acción Regional de Primatologia para las Américas" por el Biólogo Filomeno Encarnación, Asesor Temporareo de la OPS/OMS.

Luego, la Comisión de Estatutos y Reglamentos de la CORP hizo las propuestas de cambio a la asamblea sobre los proyectos estudiados, las que fueron aprobadas. Los Estatutos y Reglamentos son parte de este Informe Final, como Anexo No. 3. Posteriormente, la asamblea aprobó por unanimidad el objetivo general y las estrategias del Plan de Acción Regional de Primatología para las Américas. El presidente solicitó a las comisiones de movilización de recursos, investigación y conservación que informarán de su labor, y las recomendaciones emanadas de ellas fueron aprobadas.

La asamblea por unanimidad aprobó un voto de aplauso a Battelle Pacific Northwest Laboratories por las excelentes facilidades brindadas para llevar a cabo esta Reunión.

En la sesión de clausura hicieron uso de la palabra el presidente de la Reunión, Ing. Manuel Uceda y el Dr. Primo Arámbulo III de la OPS/OMS.

RECOMENDACIONES APROBADAS

La I Reunión del Comité Regional aprobó las siguientes recomendaciones:

A LOS PAÍSES:

1. Pedir a los países desarrollados que movilicen recursos económicos hacia los países con poblaciones nativas de primates no humanos, a fin de que estos desarrollen programas de conservación y uso sostenible de primates para sus investigaciones biomédicas.

2. Instar a los países, que aún no lo han hecho, establecer y/o fortalecer sus proyectos nacionales de conservación de primates no humanos en su habitat natural, en cautiverio o semicautiverio.

3. Solicitar a los países que establezcan acuerdos bilaterales para desarrollar actividades conjuntas que permitan la conservación de recursos naturales con énfasis en las poblaciones de primates no humanos.

4. Solicitar a los países que en lo posible adopten o adapten las estrategias y acciones del Plan de Acción Regional de Primatología para las Américas en sus planes y programas nacionales.

A LA ORGANIZACIÓN PANAMERICANA DE LA SALUD (OPS/OMS):

1. Pedir a la OPS/OMS que facilite y coordine la transferencia de tecnología relacionada a la primatología especialmente a los países miembros del CORP.

2. Pedir a la OPS/OMS que formule las actividades de cooperación técnica en base al objetivo general y estrategias del Plan de Acción Regional de Primatología para las Américas aprobadas en esta I Reunión del CORP.

3. Solicitar a la OPS/OMS que siga cooperando en la búsqueda de recursos económicos y técnicos para el desarrollo de la primatologia en los países de Región.

4. Solicitar a la OPS/OMS que actúe como secretaría ex-officio de los acuerdos interpaíses a que hace referencia la recomendación 3 a los países.

5. Pedir a la OPS/OMS que solicite a la Fundación Panamericana para la Salud y Educación (PAHEF) establezca un proyecto donde se depositen las subvenciones y donaciones, entre otras, de personas y/o instituciones oficiales y privadas para el desarrollo de la primatología y de actividades del CORP.

Recomendaciones de las Comisiones
de Investigación, Conservación y Movilización de Recursos

Comisión de Investigación:

1. Estimular investigaciones tendientes a optimizar la cría de monos platirrinos en cautiverio en los países del neotrópico.

2. Estimular investigaciones sobre la historia natural de los primates neotropicales, tendientes a promover su conservación en los ambientes silvestres.

3. Formar recursos humanos para la investigación básica y aplicada en los países del neotrópico.

4. Establecer en cada institución o universidad dedicada al estudio de la primatología, un centro de documentación y/o referencia a través de bibliografía actualizada y conexión con las bases de datos biológicos, especialmente primatológicos.

Esta comisión está presidida por el Dr. Julio C. Ruiz de Argentina e integrada por la Dra. María Rovira de Paraguay, el Dr. Milton T. de Mello de Brasil y el Dr. Alejandro Estrada de México.

Comisión de Conservación:

1. Posibilitar la participación del CORP en el desarrollo del Plan de Acción para los neotrópicos adelantado por el grupo de especialistas de primates del UICN.

2. Orientar recursos provenientes de instituciones que utilizan primates para biomedicina hacia la identificación de nuevas áreas de conservación y/o implementación de las ya existentes.

3. Condicionar el uso de primates no humanos a partir de la producción de individuos en condiciones naturales o controladas, a la inversión de recursos económicos comparables a los costos reales de producción y a los precios de mercado internacional.

4. Incluir los objetivos de los programas nacionales de primatología dentro de otros convenios multi y/o binacionales como el Tratado de Cooperación Amazónica.

5. Iniciar procesos de capacitación a corto y largo plazo sobre los diferentes aspectos relacionados con el manejo de primates en ambientes naturales y/o controlados.

6. Promocionar el conocimiento mediante métodos prácticos de la diversidad de primates de cada región neotropical, que incluyan además de la información relativa a cada especie, mensajes sobre su importancia y conservación.

7. Desarrollar un banco de datos sobre el estado de conservación y manejo a través de organizaciones de reconocido prestigio internacional que incluya además información sobre las actividades de cría en los países de origen y de destino.

8. Estimular la unificación de criterios en sistemática que permitan unificar el conocimiento sobre la presencia de especies y distribución de las mismas.

Esta comisión esta presidida por el Dr. José Vicente Rodríguez de Colombia e integrada por el Dr. Jordán Paulo Wallauer de Brasil ; Dr. Alejandro Estrada de México, el Biólogo Filomeno Encarnación de Perú y el Dr. Roderic Mast de Conservación Internacional.

Comisión de Movilización de Recursos:

La responsabilidad de los países miembros es remitir al secretariado un documento definiendo sus fuentes e identificando sus objetivos para el 31 de abril de 1991.

Para definir las fuentes los países miembros han sido divididos en aquellos que tienen especies nativas e.i. Perú y aquellos sin primates nativos e.i. Estados Unidos de América. El siguiente lineamiento será de utilidad para completar el documento.

Países con Poblaciones Nativas:	1)	Número de especies y un estimado de poblaciones relativas de cada especie.
	2)	Identificar Centro de Primates existentes y otros tipos de instituciones involucradas en la conservación y producción de primates.
	3)	Descripción de las facilidades, incluyendo personal, tiempo de financiamiento, filiación y otros datos pertinentes.
	4)	Dependencias de Gobierno responsables de la autorización para la exportación de primates no humanos.
Países sin Población de Primates Nativos:	1)	Destreza y experiencia en colonias de crianza de primates no humanos en cautiverio, sanidad y servicios de diagnóstico. Organización de talleres de trabajo para transferir información a los países miembros.

2) Relación de investigadores que requieren primates no humanos en los países de origen o para la exportación a instituciones de investigación.

3) Facilitar información a la comunidad científica de la disponibilidad de primates no humanos.

4) Relación de instituciones que tienen o hacen uso de primates neotropicales en la investigación, pruebas diagnósticas, docencia y parques zoológicos.

5) Relación de los requerimientos de cada especie neotropical usados en apoyo a esos programas.

Los objetivos son comunes para todo los países miembros, por lo que cada país miembro incluirá en el documento las estrategias por los cuales estos objetivos serán logrados.

1. Intercambio de información e investigación colaborativa entre los países miembros.

2. Desarrollo de infraestructura para asegurar la conservación de las especies y manteniendo del potencial de investigación de los países de origen.

3. Preservar y estimular el desarrollo económico de las fuentes de primates en los países de origen mediante el establecimiento de programas de educación ambiental y conservación.

Esta comisión está presidida por el Dr. James Wong de Canadá e integrada por el Sr. Jean Baulu, de Barbados y el Dr. Stephen Potkay, de Estados Unidos de América.

EN FE DE LO CUAL, el Presidente de la Reunión y el Secretario ex officio, firman el presente Informe Final en los idiomas español inglés, cuyos textos tendrán igual valor auténtico.

HECHO en la ciudad de Seattle, Estado de Washington, Estados Unidos de América el día treintiuno de octubre de mil novecientos noventa. El Secretario depositará los textos originales en los archivos de la Oficina Sanitaria Panamericana y enviará copias a los miembros del Comité y a los Gobiernos Miembros de la Organización.

Ing. Manuel Uceda
Director General de Forestal y Fauna
Ministerio de Agricultura del Perú
Presidente

Dr. Primo Arambulo III
Secretario ex-officio
del Comité Regional de
Primatología para las Américas
Oficina Sanitaria Panamericana

Conservación de los Primates de los Neotrópicos

Russell A. Mittermeier
Departamento de Ciencias Anatómicas, HSC
Universidad Estatal de Nueva York
Stony Brook, NY 11794
EE.UU.

y

Conservation International
1015 18th Street NW
Washington, D.C. 20036
EE.UU.

Warren G. Kinzey
Departamento de Antropología
City College y Graduate Center
Universidad de la Ciudad de Nueva York
Convent Avenue, E. 138th Street
Nueva York, NY 10031
EE.UU.

Roderic B. Mast
Conservation International
1015 18th Street, N.W.
Washington, D.C. 20036
EE.UU.

La mayor amenaza para los primates de los neotrópicos es la pérdida de selva tropical para dedicarla a actividades agrícolas, ganaderas, madereras e hidroeléctricas. La caza de los primates para el consumo o venta y la captura de animales vivos para el comercio de animales domesticables o mascotas y para fines biomédicos también representa una amenaza para ciertas especies. Veintiocho de las 76 especies de primates de los neotrópicos se consideran en peligro de extinción y la mayor densidad de grupos taxonómicos en peligro ocurre en la selva de la costa atlántica del Brasil, seguida de la Amazonía y Mesoamérica. Los géneros de primates de los neotrópicos que están más en peligro son los *Brachyteles* y *Leontopithecus*, seguidos de algunas especies de *Ateles, Lagothrix, Saguinus*, y el tití o frailecillo centroamericano (*Saimiri oerstedii citrinellus*). Se analizan soluciones para los problemas de la conservación de los primates, entre ellas la conservación del habitat, la protección contra la caza furtiva y el comercio internacional de primates, la educación, la investigación y la reproducción en cautiverio, y se proporcionan ejemplos de actividades eficaces de conservación.

Los problemas

Las principales amenazas para los primates silvestres en todo el mundo pueden clasificarse como presión directa contra los propios animales o como destrucción y alteración de su habitat. La magnitud relativa de estos dos peligros varía entre las especies y regiones, pero uno o ambos influyen prácticamente en todas las poblaciones existentes de primates.

Amenazas para el habitat

El factor más importante contribuyente al descenso en el número de primates no humanos es la destrucción del habitat (Mittermeier & Cheney, 1987) debido principalmente al desarrollo económico de los ambientes naturales (Myers, 1976). Más del 90 por ciento de todas las especies de primates ocurren en selvas tropicales y, a medida que desaparecen estas selvas, así lo hacen también los animales que dependen de ellas para su supervivencia.

La causa última de la destrucción del habitat es la explosión de la población humana, que es mayor en los países pobres y subdesarrollados de los trópicos. La adición de millones de personas desposeídas de tierras cada año a los países en desarrollo ya sobrecargados hace que la tendencia hacia una pérdida de selvas en gran escala parezca irreversible (Mittermeier & Cheney, 1987). Las tasas de crecimiento de la población en casi todos los países tropicales son mucho más elevadas que en el mundo desarrollado. La tasa de crecimiento anual de la población en los Estados Unidos es de 0,86 por ciento y las tasas descendieron realmente por debajo de cero en algunos países europeos; en fuerte contraste, la tasa anual de crecimiento del Brasil es de 2,07 por ciento, la de Indonesia, de 1,74 por ciento, la de Zaire de 3,04 por ciento y la del Perú de 2,51 por ciento (Instituto Internacional del Ambiente y Desarrollo e Instituto de Recursos Mundiales, 1987) y éstos figuran entre algunos de los principales países del mundo en lo que respecta a la diversidad de los primates y la cubierta forestal tropical (Mittermeier, 1987d). Aunque es inevitable que desaparezca parte de las selvas y de los primates del mundo, el papel de los defensores del medio ambiente ha de ser el limitar esta pérdida en el mayor grado posible.

Las selvas neotropicales se están destruyendo por distintas razones: agricultura, ganadería, demanda de carbón vegetal y leña para combustible, tala comercial de madera dura, construcción vial y proyectos hidroeléctricos masivos. Las estimaciones recientes, aunque difieren algo, están generalmente de acuerdo en que, a fines de 1970, se estaban destruyendo como mínimo 0,6 por ciento por año de todas las selvas tropicales cerradas de hoja ancha (Lanley, 1982; Melillo y colaboradores, 1985) y esta cifra es ahora más elevada sin duda dado el constante incremento de las actividades de desarrollo que ha tenido lugar en los trópicos.

El grado y efecto de la destrucción del habitat varía de una región a otra, aunque las estimaciones son aproximadas. En Bolivia, sólo 0,2 por ciento de las selvas se están destruyendo anualmente pero, en las selvas de la costa atlántica del Brasil, tanta selva ha sido destruida ya que cualquier pérdida adicional de la selva constituye una grave amenaza para las poblaciones restantes de primates, muchas de las cuales están ya en peligro de extinción. Análogamente, si continúan las tendencias actuales, muchas de las selvas de América Central también desaparecerán a principios de la década de 1990 (Myers, 1984).

La rápida tasa de destrucción de la selva en algunas zonas subraya la vulnerabilidad de las regiones que aún incluyen grandes extensiones de selva. Por ejemplo, desde la llegada de los europeos, las selvas de la costa atlántica en el este del Brasil han sido reducidas a 1-5 por ciento de su extensión original. Puesto que la mayor parte de esta destrucción ha ocurrido en los últimos 20 ó 30 años, es fácil imaginar cómo grandes extensiones de selva, tales como las que se hallan en la Amazonía, pudieran también ser vulnerables a una rápida destrucción (Mittermeier y colaboradores, 1982). Si continúan las tendencias actuales, se estima que, para el año 2020, las selvas de Amazonía habrán quedado reducidas en un 50 por ciento y que todas las demás selvas accesibles fuera de los límites de los parques y las reservas habrán quedado totalmente destruidas (Myers, 1987). Otras estimaciones indican que la destrucción de la Amazonía pudiera ser aún más rápida y que la mayor parte de la selva desaparecerá para fines de siglo o antes (Fearnside, 1982).

Agricultura. La causa individual más grave de destrucción de la selva en todo el mundo es la conversión a la agricultura, bien a la agricultura tradicional de subsistencia mediante tala y quema o al cultivo comercial en gran escala. En zonas fértiles, tales como en la región oriental del Brasil, la mayor parte de la selva ha sido sustituida ya por granjas o plantaciones permanentes (Mittermeier y colaboradores, 1982). En áreas de baja fertilidad del suelo (por ejemplo, en gran parte de la Amazonía) la tierra puede soportar únicamente unas cuantas estaciones de siembra intensiva antes de que la erosión y el uso excesivo agoten sus elementos nutritivos. Si las zonas desbrozadas son pequeñas y se dejan sin laborear por cierto tiempo, dicha agricultura puede ser ecológicamente beneficiosa, pero la tasa de crecimiento de la población supera en mucho al bajo rendimiento de estos métodos cuidadosos. El resultado es el de que las parcelas degradadas se abandonan y se desbrozan nuevas zonas de selva para iniciar de nuevo el ciclo de destrucción (Departamento de Estado, 1980).

Explotaciones ganaderas. La segunda causa importante de destrucción de las selvas en América Central y del Sur es la explotación ganadera. En contraste con la agricultura de subsistencia, la explotación ganadera es una actividad en la que participan a menudo corporaciones extranjeras y recibe un apoyo activo del gobierno. El Brasil, por ejemplo, promueve la explotación ganadera en la Amazonía como parte de su campaña de ampliar el control sobre el interior y desarrollar su mercado de exportaciones. Entre los grandes consorcios ganaderos en el Brasil figuran corporaciones multinacionales tales como la Volkswagen Corporation, que tiene una concesión de 1.400 kilómetros cuadrados de selva amazónica (Myers, 1984). Irónicamente, las explotación ganadera en las zonas de selvas tropicales es mucho menos eficiente que los pastizales subtropicales existentes, pero las condiciones ofrecidas por el gobierno para fomentar el desarrollo de la Amazonía son muy atractivas; por tanto, existe poco aliciente para adoptar mejores prácticas de utilización racional de la tierra (Myers, 1984).

Arboles para combustible y carbón vegetal.

La corta de árboles para combustible también es una amenaza importante y le corresponde un 80 por ciento de toda la madera extraída de la selva tropical. De ordinario, se cree que la necesidad de leña y carbón vegetal surte sólo un efecto local, pero también hay un extenso comercio internacional de carbón vegetal y éste está ocasionando una importante destrucción mundial de selvas tropicales (Oficina de Evaluación Tecnológica, 1984). En algunos países, el uso de combustible es la causa principal de la desforestación [por ejemplo, partes de América Central (Mayers, 1984)]. Se cree que, al menos 1,5 mil millones de personas que viven en el Tercer Mundo se ven afectadas por escasez de leña, y las selvas están sufriendo tasas crecientes de recolección de madera (Myers 1984). En toda la zona oriental del Brasil, están generalizados los hornos en los que se convierten a carbón vegetal los árboles nativos y eucaliptos introducidos.

Tala comercial. La tala industrial, tanto para uso local como para exportación, produce un 20 por ciento del volumen total de madera extraída de las selvas tropicales. Dicha tala agota los recursos de la población local en tanto que, en el caso de la exportación, los principales beneficios van a consumidores extranjeros, a gobiernos de otros países o compañías grandes. El Japón depende en particular de las fuentes exteriores de pulpa y utiliza cada vez más los árboles de madera dura sudamericanos (Myers, 1984). Aunque la extracción comercial de madera es generalmente selectiva, aún tiene un efecto adverso sobre las especies de primates tales como el *Ateles paniscus* y el *Lagothrix lagothricha* que dependen de selvas primarias maduras (Wolfheim, 1983; Skorupa, 1986). Además, hay pruebas crecientes de que la productividad de los árboles residuales queda gravemente reducida con la tala selectiva (Skorupa, 1986). Tristemente, aunque a las compañías sólo se les otorga de ordinario licencias para extraer una pequeña porción de una selva, la corrupción local y la falta de aplicación de la ley resultan con frecuencia en el abuso de los contratos. Finalmente, los caminos forestales construidos para sacar la madera de la selva producen a veces más daño que la propia eliminación de los árboles.

Proyectos hidroeléctricos. Los proyectos hidroeléctricos destruyen a menudo algunas de las zonas de selvas neotropicales más importantes y menos perturbadas, incluidas las que se encuentran en parques o reservas. Dichas zonas "protegidas" son seleccionadas a menudo como lugares para proyectos debido a que están relativamente deshabitadas y a que ya están en manos del gobierno. En algunos casos, el tamaño de la zona inundada puede ser bastante extenso. El Proyecto de Tucurui en la región del Amazonas del Brasil, por ejemplo, inundó unos 3.000 km. cuadrados (A. Johns, comunicación personal) y también se tienen programados para la región otros proyectos de importancia comparable.

Amenazas directas para los animales

Caza. Los primates son cazados por distintas razones, pero principalmente como fuente de alimentos. Al igual que ocurre con la destrucción del habitat, los efectos de la caza varían considerablemente de una región a otra y entre las especies. La caza de primates para alimento presenta una amenaza importante en la región del Amazonas donde miles de primates son abatidos cada año para consumo inmediato o para venta en mercados locales. En algunos lugares (por ejemplo, en el interior de Surinam; en la región del Amazonas del Ecuador), hasta 35 por ciento de la carne consumida por las poblaciones silvícolas se deriva de los primates (Mittermeier, 1977, 1987b; Yost y Kelley, 1983). La caza de primates está a menudo prohibida por la ley, pero la ejecución de la ley es rara o inexistente en las zonas remotas donde ocurre la mayor parte de esta actividad (Mittermeier y Coimbra-Filho, 1977; Mittermeier, 1987b).

Cuando la caza es común, no afecta a todas las especies por igual. Los cazadores del Amazonas prefieren especies de simios de cuerpo grande y cola prensil (Peres, 1987) y los choros (*Lagothrix lagothricha)*, los frailecitos (*Saimiri sciureus*) y, en menor grado, los monos aulladores y los monos negros (*Cebus apella*) son extensamente perseguidos en la mayor parte del Amazonas. Estas especies se han extinguido localmente en muchas partes del territorio que antes ocupaban, incluso en habitat de selvas primarias por lo demás apropiados para su supervivencia (Mittermeier y Coimbra-Filho, 1977; Soini, 1982; Mittermeier, 1987b). Los primates más pequeños, tales como los frailecitos y los pichicos, rara vez son perseguidos ya que apenas si ofrecen recompensa al cazador por el cartucho gastado.

Los primates también son destruidos para obtener la piel u otras partes del cuerpo para fines ornamentales. En la región del Amazonas, por ejemplo, varias tribus utilizan los dientes, cráneos y pieles de monos para ornamentación y estos artículos se venden ocasionalmente en las tiendas de turistas. Las tupidas colas de los huapos (género *Pithecia*), son populares como plumeros para quitar el polvo, y las pieles de monos, gorros de piel de mono y monos disecados se ven con frecuencia en las tiendas de curiosidades de las principales ciudades de la región del Amazonas. Sin embargo, principalmente, este tipo de utilización de los primates es un subproducto de la caza para carne.

Finalmente, algunos primates neotropicales son cazados como enemigos de la agricultura. Aunque las especies semiterrestres del Viejo Mundo tales como los babuinos o mandriles, los macacos y los micos son más destructivos para las plantaciones agrícolas, los primates neotropicales, especialmente los *Cebus*, destruyen cultivos en muchas partes de Sudamérica. Ciertamente, los nombres populares para *Cebus nigrivittatus* en Surinam (*nyangkaru mongi)* y para *Cebus apella* en Colombia (*maicero)* se traducen como "comedores de maíz" (Mittermeier, 1987b). La destrucción de los primates como supuestos enemigos de los cultivos puede producir pérdidas mayores de las previstas. A medida que continúa la usurpación del habitat de los primates, las especies más adaptables probablemente dependerán más de los cultivos del hombre. El resultado desafortunado será un mayor conflicto entre el ser humano y los primates no humanos lo que, a su vez, exacerbará la dificultad de obtener un apoyo verdadero para la conservación.

Captura de animales vivos. La segunda amenaza directa para los primates es la captura de animales vivos para el comercio como mascotas y para fines de investigación biomédica. En el punto culminante del comercio internacional de primates, durante los años 50 y 60, la exportación de primates de América Central y del Sur para estos fines llegó a cientos de millares de animales (Mack y Mittermeier 1984). Las especies que han sufrido más por el comercio internacional son, entre otras el frailecito, exportado de Colombia y Perú para el comercio de mascotas o animales domesticables; los *Saguinus oedipus* del norte de Colombia, capturados para los laboratorios de investigación (Hernández- Camacho y Cooper, 1976); los choros y los monos araña que son aún mascotas populares en la región del Amazonas (Mittermeier y Coimbra-Filho, 1977) y el *Leontopithecus chrysomelas*, una forma especialmente en peligro de extinción que en fecha reciente ha entrado en el comercio ilegal tanto en el Brasil como internacionalmente (Mallinson, 1984).

El comercio local en primates dentro de los países de origen también es importante en potencia. Algunos países en desarrollo tales como el Brasil están comenzando a utilizar especies autóctonas en sus propios programas de investigación biomédica y cada año se venden cientos de tities (*Callithrix* spp.) en el mercado ilegal de mascotas en las grandes ciudades de Río de Janeiro y Sao Paulo. Muchos de estos primates entran en el mercado internacional a través de los vendedores de animales domesticables o mascotas en países tales como el Japón.

En la región del Amazonas, donde los primates ya son cazados por su carne, el papel de animales domesticables o mascotas que se les da es un subproducto de la caza para alimentación. De ordinario, las crías de los primates constituyen las mejores mascotas y a menudo se obtienen dando muerte a la hembra adulta y quitándole la cría que cuelga de ella. Para los primates tales como el *Lagothrix lagothricha*, incluso el pequeño incentivo adicional del comercio de mascotas puede aumentar notablemente los problemas de conservación. Puesto que los choros son, a la vez, la fuente favorita de carne de simio y de mascotas más deseable, de ordinario un cazador disparará contra las hembras que tienen crías lactantes en preferencia a otros animales. Los efectos de esta caza selectiva sobre las tasas de reproducción de una especie lenta en reproducir tales como los choros pueden ser claramente peligrosos.

Primates Neotropicales que están más en peligro de extinción

De las 76 especies de monos del Nuevo Mundo actualmente reconocidas, 28 especies (y/o subespecies) están ya incluidas como en peligro, vulnerables, raras o indeterminadas en el Libro Rojo de Datos de IUCN (Thornback & Jenkins, 1982; véase el Cuadro 1). Esto representa únicamente una estimación mínima del número de primates neotropicales en peligro de extinción ya que hay muchas especies cuyo estado de conservación es desconocido.

Cuando se considera la distribución de los grupos taxonómicos en peligro por región, la mayor densidad (dos géneros y nueve especies) ocurre en las selvas de la costa atlántica del Brasil, seguidas de la región del Amazonas y Mesoamérica (véase Mittermeier, 1987d; Figura 1). En términos de países, el Brasil, con 19 grupos taxonómicos en peligro (18 de ellos endémicos), se halla también a la cabeza de la lista.

Los monos del Nuevo Mundo que están en peligro más crítico de extinción son los géneros *Brachyteles* y *Leontopithecus* (Cuadro N° 1). Ambos de estos géneros ocurren sólo en pequeñas extensiones residuales de la región de la selva de la costa atlántica en el Brasil oriental y están ambos al borde de la extinción. El muriqui (*Brachyteles arachnoides*) es el primate sudamericano más grande. Su población restante puede incluir hasta sólo 300-400 individuos (Mittermeier, 1987b,c). Análogamente, el género de *Leontopithecus,* incluye tres especies (Rosenberger y Coimbra-Filho, 1984) que todas están en gran peligro de extinción. Estas especies ocurren únicamente en unas cuantas extensiones de selvas residuales y las reservas creadas para ellos son, en algunos casos, pequeñas y están ocupadas parcialmente por colonos (Mittermeier y colaboradores, 1982). Las amenazas para ambos géneros se analizan en mayor nivel de detalle en Mittermeier (1987a).

Otros primates del Nuevo Mundo en grave peligro de extinción son los monos araña (*Ateles)* y los choros (*Lagothrix).* Aunque ambas especies tienen una extensión geográfica grande, son cazadas intensamente y, en la mayoría de los casos, requieren una selva relativamente intacta para su supervivencia. Además, debido a su gran tamaño, se reproducen a tasas relativamente lentas (en los *Ateles*, los intervalos entre nacimientos son de unos tres años. [Eisenberg,1979]) y, por consiguiente, se recuperan lentamente de cualquier clase de explotación. No es sorprendente que el *Ateles* y el *Lagothrix* hayan desaparecido de muchas partes de su distribución anterior, incluso en zonas de selvas primarias apropiadas (Mittermeier y Coimbra-Filho, 1977; Mittermeier, 1987b).

Aunque la mayor parte de las especies de primates neotropicales en peligro de extinción ocurre en la selva de la costa atlántica y en la región del Amazonas, entre los primates en peligro endémicos de otras regiones figuran el choro de cola amarilla (*Logothrix flavicauda)* del norte del Perú, el frailecito centroamericano (*Saimiri oerstedii*) y el pichico (*Sanguinus oedipus geoffroyi)* de Panamá. El choro de cola amarilla (*Logothrix flavicauda*), la única especie de primates andinos, está limitado a una pequeña zona en el norte del Perú. Este mono, que se volvió a descubrir en 1974 después de un lapso de casi 50 años (Mittermeier y colaboradores, 1975, 1977), se ve amenazado por la destrucción del habitat y por la caza en su habitat de la selva de zona de montaña elevada que con anterioridad se hallaba aislada.

El frailecito centroamericano se halla sólo en la región occidental de Panamá y en porciones adyacentes de la región sudoriental de Costa Rica. Aunque los frailecitos figuran de ordinario entre los primates neotropicales más abundantes, el *S. oerstedii* se considera en gran peligro de extinción. Un reciente estudio ecológico y encuesta de población de simios en Costa Rica puso de manifiesto que no hay más de 3.000 individuos supervivientes (Boinski, 1986) y es posible que estos animales costarricenses sean los únicos que aún sobreviven. Afortunadamente, cada una de las dos subespecies

Cuadro Nº 1. Grupos taxonómicos de primates neotropicales que están más en peligro de extinción.

Géneros en peligro

Brachyteles
Brachyteles arachnoides
Leontopithecus
Leontopithecus rosalia
Leontopithecus chrysomelas
Leontopithecus chrysopygus

Especie en peligro	**Subespecies en peligro**
Callithrix aurita	*Callithrix argentata leucippe*
Callithrix flaviceps	
Callithrix geoffroyi	*Saguinus bicolor bicolor*
Saguinus oedipus	*Chiropotes satanas satanas*
Callicebus personatus	*Cacajao calvus calvus*
C. p. personatus	
C. p. nigrifrons	*Cebus apella xanthosternos*
C. p. melanochir	*Alouatta belzebul ululata*
Saimiri oerstedii	*Ateles geoffroyi azuerensis*
S. o. oerstedii	*Ateles geoffroyi panamensis*
S. o. citrinellus	*Ateles geuffroyi grisescens*
Alouatta fusca	*Ateles fusciceps fusciceps*
A. f. fusca	
A. f. clamitans	*Ateles belzebuth hybridus*
	Ateles belzebuth marginatus
Lagothrix flavicauda	

está representada en uno de los parques nacionales de Costa Rica, el *S. oestedii* en el Parque Nacional de Corcovado, y el *S. o. citrinellus* en el Parque Nacional de Manuel Antonio, aunque en este último caso las estimaciones de la población del primate son bajas y el habitat está gravemente degradado.

El *Saguinus oedipus* es otro mono en grave peligro de extinción. Se halla únicamente en la región noroccidental de Colombia, un área que ha estado sometida a grave destrucción de la selva durante las dos últimas décadas. Muchos de estos simios fueron capturados para el comercio de primates en los años 60 y principios de los 70. Se estima que se exportaron unos 20 a 30.000 animales durante los años de máximo comercio de simios (Hernández- Camacho y Cooper, 1976). El *S. oedipus* está bien representado en cautiverio, con unos 1.161 animales en zoológicos y laboratorios de investigación en los Estados Unidos y Europa. En fecha reciente se ha terminado un libro de estudio sobre esta especie (Tardiff y Colley, 1987).

Existe poca información actualizada sobre el estado del *S. oedipus* en libertad. Sin embargo, esta especie había quedado ya diezmada cuando se realizó el último estudio en 1975 (Neyman, 1977). Las observaciones preliminares de un estudio reciente indican que la situación se ha deteriorado desde entonces (V. Rodríguez y J. Hernández-Camacho, comunicación personal). Es probable que estos monos se estén aproximando a la situación de los *Leontopithecus* en el Brasil.

Soluciones Posibles

Entre las soluciones posibles para los problemas de la conservación de los primates en los neotrópicos (Figuras N° 1 y 2) figuran la conservación del habitat, con protección contra la caza furtiva y la captura incontrolada de animales vivos, actividades de educación y toma de conciencia del público, capacitación (especialmente de expertos locales en conservación), investigación de poblaciones en libertad, propagación de animales en cautiverio y reintroducción a ambientes naturales donde sea apropiado.

Conservación del habitat

Lamentablemente, la conservación de las selvas tiene aún una baja prioridad económica o política debido a que ésta no se traduce fácilmente a beneficios inmediatos. En el presente, sólo una fracción de los fondos proporcionados anualmente por instituciones tales como el Banco Mundial y la Agencia de los Estados Unidos para el Desarrollo Internacional (USAID) se dedica a conservación de las selvas tropicales. La replantación y utilización racional de árboles silvícolas valiosos pudiera hacer posible la utilización sostenida de las selvas, pero, en la práctica, estas actividades son raras y en pequeña escala (Departamento de Estado de los Estados Unidos, 1980; Oficina de Evaluación Tecnológica, 1984). Se necesitan mayores esfuerzos para administrar debidamente los parques y reservas y crear nuevos parques y reservas donde sea necesario. También es importante hallar formas de hacer que los parques se sustenten a sí mismos económicamente y, quizás que se hagan rentables, puesto que esto animará a los gobiernos locales y nacionales a apoyar una creación y mantenimiento más activos de los parques.

Selvas de la costa atlántica. El área de mayor prioridad en los neotrópicos es evidentemente la región de selvas de la costa atlántica en el Brasil oriental, con dos géneros endémicos y nueve especies endémicas de primates en peligro de extinción. Esta región es el corazón agrícola e industrial del país así como su centro de población. Por consiguiente, sólo de 1 a 5 por ciento, aproximadamente, de su cubierta original de selva (estimada originalmente en más de 1.000.000 de km cuadrados), queda ahora intacta.

FIGURA Nº 1. Número de especies de primates neotropicales y géneros en cada región fitogeográfica principal.

FIGURA N° 2. Especies y géneros de primates neotropicales por país.

Desde 1979, el World Wildlife Fund (WWF; Fondo de Fauna Mundial) ha apoyado un amplio programa dedicado a la región de la selva de la costa atlántica, con numerosos proyectos que abordan distintos aspectos de la conservación. Todo el programa del WWF fue el resultado de un pequeño estudio de los primates iniciado en 1979, y la conservación de los primates sigue siendo un componente importante de este programa de la selva de la región atlántica. El grueso de esta labor la está realizando un equipo de unos 20 estudiantes e investigadores brasileños, con alguna participación norteamericana y británica (Mittermeier, 1987a)

Región del Amazonas. El área que sigue en prioridad es la región del Amazonas. Este área es, en mucho, la mayor región de selvas tropicales del mundo y el lugar donde viven la mayoría de las especies de simios del Nuevo Mundo (Mittermeier y Cheney, 1987) y hay en ella hasta 13 especies originarias en algunas zonas.

Al nivel genérico, los primates de mayor preocupación en la región del Amazonas son los *Ateles* y los *Lagothrix*. Ninguno de éstos se hallan en una situación tan crítica como los *Brachyteles* y los *Leontopithecus* que están en peligro de convertirse en géneros extintos, pero tanto los monos araña como los choros están desapareciendo con rapidez y muchos han desaparecido ya de gran parte de la región del Amazonas. La especie siguiente de monos del Amazonas en peligro de extinción está constituida por varias subespecies con habitat muy limitados en áreas de fuerte desarrollo. Están incluidas en este grupo las dos subespecies meridionales del huapo barbudo (*Chiropotes satanas satanas* y *C. s. utahickii)*, el huapo blanco (*Cacajao calvus calvus)*, el pichico bicolor (*Saguinus bicolor bicolor)* y el tití blanco (*Callithrix argentata leucippe*). Todos estos cuatro animales son endémicos en la región del Amazonas del Brasil (Rylands y Mittermeier, 1982).

Aunque en menor grado que las selvas de la costa atlántica y el Amazonas, las selvas de Centroamérica, las selvas de montaña alta del norte del Perú y Colombia y las selvas más secas de la Argentina, Paraguay, Bolivia también han de ser consideradas como áreas críticas para la conservación de los primates. La estrategia para estas regiones debería concentrarse en proteger especies en gran peligro de extinción (por ejemplo, los choros de cola amarilla y el *Saguinus oedipus*), y en establecer y mantener zonas protegidas apropiadas cada vez que sea posible.

Protección de los propios primates

Aunque la protección del habitat es un objetivo principal de la conservación de los primates neotropicales, también es necesario asegurar que los propios animales reciben protección. Esto significa poner fin a todo el tráfico ilegal y, por lo demás, destructivo de primates, tanto en el país de origen como en los países importadores.

El comercio de primates silvestres sólo puede reducirse si los países consumidores se niegan a adquirirlos. Los incentivos económicos por la captura de primates son elevados y seguirán siéndolo en tanto que los individuos, los zoológicos y las instituciones médicas estén dispuestas a participar en el comercio de primates vivos. El descenso en el comercio de mascotas para los primates neotrópicos durante la última década puede atribuirse principalmente a la imposición de prohibiciones de exportación por los países de origen, la aplicación de restricciones de importación por los países usuarios y una disminución en la demanda de primates para la investigación biomédica (Mack y Mittermeier, 1984).

El comercio internacional de primates se ha reducido notablemente desde la redacción de la Convención sobre el Comercio Internacional de Especies en Peligro de Flora y Fauna (CITES) en 1973 y la imposición de cierto número de prohibiciones a la exportación por los países de origen. Las partes firmantes de CITES convinieron en prohibir el intercambio comercial de especies en peligro y en vigilar el comercio de otras especies que pueden verse en peligro de extinción. Hasta la fecha, todos los países latinoamericanos excepto México han ratificado o accedido a CITES. Aunque la convención proporciona pautas valiosas para el control del comercio de la fauna, sólo es eficaz en el grado en que los países la apliquen, y algunos países siguen sus requisitos más estrictamente que otros. Las excepciones que permiten, por ejemplo, la venta legal de animales nacidos en cautiverio o la importación de animales capturados en países específicos contribuyen a formas tradicionales de comercio ilegal, sobornos, corrupción y a la identificación equívoca intencionada de las especies (Wolfheim, 1983).

Aunque de vez en cuando se viola la convención CITES, ésta ha producido una reducción considerable en el número de primates no humanos importados a los países desarrollados cada año. Por ejemplo, en 1968, antes de la ratificación del acuerdo CITES, los Estados Unidos importaron 113.714 monos y simios; para 1983, este número había descendido a 13.148 (Wolfheim, 1983; Mack y Mittermeier, 1984). También se han hecho esfuerzos por establecer pautas para el uso de los primates en peligro de extinción y capturados en su estado salvaje para fines de investigación biomédica. En 1981, el Grupo de Especialistas en Primates de IUCN formuló una política sobre el uso de los primates en la investigación biomédica que recomendó que las especies en peligro, vulnerables y raras sólo se utilicen para la investigación médica cuando se obtienen de colonias existentes, en cautiverio, que procrean independientemente. Es importante que las instituciones de investigación que utilizan a primates tomen conciencia del estado de conservación de la especie que utilizan, que empleen primates lo más prudentemente que sea posible y que hagan todos los esfuerzos necesarios para criar los primates que necesiten.

Educación y toma de conciencia por el público

Para que cualquier actividad de conservación de primates tenga éxito, ha de haber una conciencia pública de la necesidad de conservación y de la importancia de los primates como un legado nacional y como un recurso, especialmente en los países en que estos ocurren. Si deseamos salvar el habitat de las selvas tropicales, nuestros esfuerzos han de dirigirse tanto a la educación como a la protección de los animales y sus habitat; tanto a crear incentivos económicos como a desarrollar parques y reservas; y a aumentar la comprensión de que la agricultura, los suministros confiables de agua y la energía hidroeléctrica dependen todos ellos últimamente de la conservación de las selvas.

En el Brasil, Perú y Colombia se están llevando a cabo varias campañas eficaces de educación. El muriqui es el mamífero endémico mayor del Brasil y, como tal, es el más apropiado como símbolo para un programa de conservación de la selva de la región atlántica y para el movimiento de conservación del Brasil en su conjunto. A fin de divulgar la situación en que se encuentra esta especie y su habitat, el Fondo de Fauna Mundial lanzó una campaña a favor del muriqui en 1981 en colaboración con la Fundación de Conservación del Brasil, el Centro de Primates de Rio de Janeiro y la Universidad de Minas Gerais. La campaña incluye programas de conferencias en las ciudades y en el interior, la preparación de exposiciones de museo y una película titulada "El Grito del Muriqui", y la distribución de camisetas, carteles, pegatinas y otros materiales educativos.

La campaña ha logrado obtener una gran cantidad de cobertura de los medios de información para el muriqui. La película ya se ha proyectado cientos de veces en el Brasil, especialmente en las pequeñas poblaciones y poblados cerca del habitat de este primate. Toda esta actividad ha conducido a un incremento general en la conciencia acerca de la conservación en el Brasil, que se ha manifestado por la aparición del muriqui en la portada de un guía de teléfonos del Brasil, en sellos de correos y en frecuentes artículos en revistas y periódicos.

En el Perú, están en vías de realización varios proyectos a favor del choro de cola amarilla (*Logothrix flavicauda*) en gran peligro de extinción. Un investigador peruano está realizando ahora una campaña de educación y toma de conciencia pública que entraña el uso de la misma clase de materiales empleados en el Brasil. Además del programa de educación, también se están investigando dos lugares posibles para nuevas áreas protegidas que sirvan de habitat al *L. flavicauda* .

Investigación de campo

La investigación científica también hace importantes contribuciones a la conservación, tales como la determinación de los lugares en los que ocurren los primates y en qué densidades, y la determinación de las dimensiones de las reservas que se requieren a fin de mantener poblaciones reproductoras viables. Los estudios del comportamiento son importantes para determinar los parámetros reproductivos y sociales que son necesarios para una buena procreación en cautiverio. La capacidad de establecer programas de propagación para especies tales como el muriqui se obtendrá últimamente de estudios de campo de su comportamiento. La investigación de campo ecológica y de comportamiento también proporciona una presencia científica que, de por sí, ofrece un grado de protección y prestigio a un área determinada. En algunos casos, una presencia de investigación científica ha proporcionado un incentivo para el establecimiento de parques y reservas nacionales donde no existían antes, tal como la reserva de propiedad privada de Fazenda Montes Claros en Minas Gerais, Brasil, que sustenta una de las poblaciones más importantes de *Brachyteles*.

Procreación en cautiverio y reintroducción en el habitat natural

Para ciertas especies en peligro de extinción, los programas de procreación en cautiverio pueden ser un componente importante de un plan general destinado a salvar a un primate en peligro de extinción. Aunque los programas de procreación ofrecen la esperanza más importante para algunas especies, éstos no deberían sustituir ni incluso considerarse como un equivalente de las actividades destinadas a conservar poblaciones de primates en su ambiente natural.

El programa de procreación en cautiverio para el *Leontopithecus rosalia* ha tenido mucho éxito y los resultados de la introducción de este animal en el ambiente natural son alentadores (Kleiman y colaboradores, 1986). Se están iniciando actividades similares de procreación en cautiverio con el *Leontopithecus chrysopygus* en peligro de extinción (Padua, 1986), el *Leontopithecus chrysomelas* y varias otras especies de la selva de la región atlántica en peligro de extinción en el Centro de Primates de Rio de Janeiro y en otros lugares.

Planes para el futuro

Es importante destacar que las actividades de conservación como las citadas arriba no se realizan en aislamiento sino, más bien, como parte un amplio esfuerzo internacional que ha sido coordinado por el Grupo de Especialistas en Primates del IUCN/SSC desde 1977. La meta de este grupo consiste en mantener la diversidad actual del Orden de los Primates, con énfasis en la supervivencia de las especies clave en peligro, dondequiera que ocurran, y en la protección de áreas con una diversidad y abundancia de primates especialmente elevadas.

Para alcanzar esta meta, el Grupo de Especialistas en Primates está formulando Planes de Acción para Conservación de los Primates para cada una de las cuatro regiones principales del mundo en las que ocurren primates: Asia, Africa, Madagascar y los Neotrópicos. El Plan de Acción para la Conservación de los Primates Africanos (Oates, 1986) y el Plan de Acción para la Conservación de los Primates Asiáticos (Eudey, 1987) están concluidos y pronto seguirán los Planes de Acción para Madagacar y los Neotrópicos.

En el Brasil, una meta se ha logrado ya en gran medida: 90 por ciento de los proyectos que apoya el Fondo de la Fauna Mundial están siendo realizados por personal brasileño y se espera que resulte una situación similar en el resto de la región de los neotrópicos dentro de los próximos diez años. Es de esperar que pueda ayudarse a todos los países de los neotrópicos y animárseles hasta el punto en que realicen sus propias actividades de conservación, se den cuenta de la extensión del problema y adquieran la capacidad económica necesaria para proporcionar soluciones. Si puede hacerse esto, entonces no hay razón para que todas las especies de primates de los neotrópicos no logren sobrevivir hasta adentrado el siglo próximo y más allá.

Referencias

Boinski, S. 1986. Evaluation of the status of the squirrel monkey (*Saimiri oerstedii*), en Costa Rica. *Prim. Conserv.* **6**, 15-16.

Eisenberg, J.F. 1979. Habitat, economy and society: Some correlation and hypothesis for the Neotropical primates, páginas 215-261. En *Primate Ecology and Human Origins*, I.S. Bernstein y E.O. Smith. eds. Nueva York, NY, USA: Garland Press.

Eudey, A. 1987. IUCN/SSC Primate Specialist Group Action for Asian Primate Conservation: 1986-1990, 41 páginas. Washington, DC, USA: World Wildlife Fund/IUCN/UNEP.

Fearnside, P.M. 1982. Deforestation in the Brazilian Amazon: How fast is it ocurring? *Interciencia* **7**, 82-88.

Hernández-Camacho J. & Cooper, R.W. 1976. The nonhuman primates of Colombia, páginas 35-69. En *Neotropical Primates. Field Studies and Conservation*, R.W. Thorington & P.G. Heltne, Eds. Washington, DC, USA: National Academy of Sciences.

Instituto Internacional para el Ambiente y el Desarrollo e Instituto de Recursos Mundiales. 1987. *World Resources 1987*. Nueva York, NY, USA: Basic Books, Inc.

Kleiman, D.G., Dietz, I.A., Dietz, J.M., Beck. B.B. & Ballon, J.D. 1986. Reintroduction of golden lion tamarins: a model for the conservation of an endangered primate. *Primate Report* **14**, 159.

Lanley, J.P. 1982. Tropical Forestry Resources. Roma, Italia: Documento de sivilcultura de la FAO No. 30.

Mack, D & Mittermeir, R.A. eds. 1984. *The Primate Trade*. Washington, DC, USA: World Wildlife Fund-US and TRAFFIC-USA.

Mallinson, J.J.C. 1984. Golden-headed lion tamarin contraband a major conservation problem. *IUCN/SSC Primate Specialist Group Newsletter* **4**, 23-25.

Melillo, J.M. Palm, C.A., Houghton, R.A. & Woodwell, G.M. 1985. A comparison of two recent estimates of disturbance in tropical forests. *Environ. Conserv.* **12**, 37-40.

Mittermeier, R.A. 1977. Distribution, synecology and conservation of Suriname monkeys, Disertación de Ph.D. en la Universidad Harvard. Cambridge, MA, USA: Harvard University.

Mittermeier, R.A. 1987a. Framework for primate conservation in the neotropical region, páginas 305-320. En *Primate Conservation in the Tropical Rain Forest*, C.W. Marsh & R.A. Mittermeier, Eds. Nueva York, NY, USA: Man R. Liss, Inc.

Mittermeier, R.A. 1977b. Effects of hunting on rain forest primates, páginas 305-320. En *Primate*

Conservation in the Tropical Rain Forest, C.W. Marsh & R.A. Mittermeier, Eds. Nueva York, NY, USA: Alan R. Liss, Inc.

Mittermeier, R.A. 1977c. Monkey in peril. *Nat. Geogr. Mag.* **171**, 387-395.

Mittermeier, R.A. 1977d. Primate conservation priorities in the Neotropical region, páginas 221-240. En *Primates: The Road to Self-Sustaining Population*, K. Benirschke, ed. Nueva York, NY, USA: Springer Verlag. Inc.

Mittermeier, R.A. & Coimbra-Fillo, A.F. 1977. Primate conservation in Brazilian Amazonia, páginas 117-166. En *Primate Conservation*, Prince Rainier of Monaco & G.H. Bourne, eds. Nueva York, NY, USA: Academic Press.

Mittermeier, R.A. & Coimbra-Fillo, A.F., Constable, I.D. Rylands, A.B. & Valle, C. 1982. Conservation of primates in the Atlantic forest region of eastern Brazil. *Int. Zoo Yearb.* **22**, 1-17.

Mittermeier, R.A. & Cheney, D.I. 1987. Conservation of primates and their habitats, páginas 477-490. En *Primate Societies*, D. I. Cheney, R.M. Seyfarth, B. Smuts, T.T. Struhsaker & R.W. Wrangham, eds. Chicago, IL, USA: University of Chicago Press.

Mittermeier, R.A. Macedo, H. de & Luscombe, B.A. 1975. A woolly monkey rediscovered in Peru, *Onyx* **13**, 41-46.

Mittermeier, R.A. Macedo, H. de & Luscombe, B.A. & Cassidy, J. 1977. Rediscovery and conservation of the Peruvian yellow-tailed woolly monkey (*Lagothrix flaricauda*), páginas 95-115. En *Primate Conservation*, Prince Ranier of Monaco & G. II. Bourne, eds. Nueva York, NY, USA: Academic Press.

Myers, N. 1976. An expanded approach to the problem of disappearing species. *Science* **193**, 198-202.

Myers, N. 1984. *The Primate Source*, Nueva York, NY, USA: Norton.

Myers, N. 1987. Trends in the destruction of rainforest, páginas 3-22. En *Primate Conservation in the Tropical Rain Forest*, C. W. Marsh & R.A. Mittermeier, eds. Nueva York, NY, USA: Alan R. Liss, Inc.

Neyman, P.F. 1977. Aspects of the ecology and social organization of free-ranging cotton-top tamarins (*Saguinus oedipus*) and the conservation status of the species, páginas 39-71. En *The biology and conservation of the Callitrichidae*, D.J. Kleiman, ed. Washington, DC, USA: Smithsonian Institution Press.

Oates, J.F. 1986. IUCNN/SSC. Primates Specialist Group Action Plan for African Primate Conservation: 1986-1990, página 44. Washington, DC, USA: World Wildlife Fund.

Padua, C.V. 1986. Captive plus wild breeding. An integrated plan for the conservation of the black lion tamarin (*Leontopithecus chrysopygus*). *Primate Report*, **14**, 158.

Peres C. 1987. Conservation of primates in western Brazilian Amazonia. A progress report to the World Wildlife Fund-US. Washington, DC, USA: World Wildlife Fund.

Rosenberger, A.I. & Coimbra-Fillo, A.F. 1984. Alorphology, taxonomic status and affinities of the lion tamarins. *Leontopithecus* (Callitrichinae: Cebidae). *Folia primatol.* **42**, 149-179.

Rylands, A.B. & Mittermeier, R.A. 1982. Conservation of primates in Brazilian Amazonia. *Int. Zoo Yearb*, **22**, 17-37.

Skorupa, J. 1986. Effects of selective timber harvesting on rain forest primates in Kibale forest, Uganda, Disertaci6n de Ph.D. en la Universidad de California, Davis. Davis, CA, USA: University of California

Soini P. 1982. Primate conservation in Peruvian Amazonia. *Int. Zoo Yearb.* **22**, 37-47.

Tardiff, S.D. & Colley, R. 1987. *International Cotton-Top Tamarin Studybook.* Oak Ridge, TN, USA: Oak Ridge Associated Universities.

Thornback, J. & Jenkins, M. 1982. *IUCN Red Data Book.* Gland, Suiza: International Union for the Conservation of Nature and Natural Resources.

U.S. Departamento de Estado. 1980. *The World's Tropical Forests: A Policy Strategy and Program for the United States*, publicaci6n No. 9117, Washington, DC, USA: Department of State.

U.S. Oficina de Evaluaci6n Tecnol6gica. 1984. *Technologies to Sustain Tropical Forest Resources*, p6gina 41. Washington, DC, USA: U.S. Congress.

Wolfheim, J.G. 1983. *Primates of the World.* Seattle, WA, USA: University of Washington Press.

Yost, J.A. & Kelley, P.M. 1983. Shotguns, blowguns and spears: the analysis of technological efficiency, p6ginas 189-224. En *Adaptive Responses of Native Amazonians*, R.B. Hames & W.T. Vickers, eds. Nueva York, NY, USA: Academic Press.

El Estado de las Poblaciones Naturales de Primates y los Beneficios de las Cosechas Sostenidas

L. Moya
F. Encarnación
R. Aquino
J. Tapia
C. Ique
P. Puertas

Centro de Reproduccion y Conservacion de Primaes No Humanos
Instituto Veterinario de Investigaciones Tropicales y de Alturas
(IVITA)
Universidad Nacional Mayor de San Marcos
Apartado Postal 621
Iquitos, Peru

I. INTRODUCCION

La importancia de los primates no humanos fue ampliamente demostrada en el mundo científico, particularmente en el campo biomédico y ecológico. En biomedicina, por el uso como modelos experimentales en beneficio de la salud humana, fisiología, biología, farmacología, psiquiatría y otras. En la ecología, cumplen un rol importante en la propagación y dispersión de semillas, participando en el mantenimiento del equilibrio y la complejidad del ecosistema tropical donde habitan.

Hasta antes de 1973, el Perú fue uno de los principales exportadores de primates sin criterios de manejo. Datos oficiales refieren la exportación entre 1964 a 1974 de 339,080 ejemplares vivos. Del total, 77,8% correspondió a *Saimiri*. Para el mismo período fue asimismo en 15 veces mas los menos sacrificados, como consecuencia de las exportaciones ilegales, mortalidad en el período de pre-exportación y caza para el consumo alimenticio en el medio rural.

A fines de 1973, el Gobierno peruano declaró la veda de la fauna silvestre Amazónica por tiempo indefinido, iniciándose simultáneamente las actividades del Proyecto Peruano de Primatología "Manuel Moro Sommo". Al presente las contribuciones son importantes para el conocimiento de la biología y ecología, así como del manejo sostenido en su hábitat natural y cautiverio, principalmente de aquellas especies de interés biomédico. Las experiencias de este Proyecto significan un importante aporte en la utilización sostenida de los primates, contempladas entre los principios de la Estrategia Mundial de la Conservación (UICN 1980).

En este documento ofrecemos aspectos del estado actual de las poblaciones naturales de los primates con énfasis de aquellas distribuidas en el Nor-Oriente peruano. También incluye los beneficios de las cosechas sostenidas de *Saguinus mystax* y *Aotus nancymai*, empleadas como modelos para las investigaciones sobre hepatitis viral y malaria respectivamente. Las experiencias han permitido valorar a las especies y sus hábitats, considerados como marginales en términos agrícolas y pecuarios tradicionales.

II. LOS BOSQUES Y LAS POBLACIONES DE PRIMATES

El bosque constituye un ecosistema caracterizado por la diversidad biológica, regulado por los componentes edáficos, los gradientes de nutrientes, la concentración de humedad edáfica y atmosférica, que responden a los factores climáticos de lluvia, temperatura, radiación solar, viento y otras. De la diversidad biológica, las especies de primates merecen especial atención por sus hábitos, patrones de conducta, organización social y uso del territorio y su distribución en las regiones biogegráficas, así como por su adaptación y evolución y afinidad morfo-anátomo-fisiológico con el hombre.

En los bosques del neotrópico, las especies de primates en su totalidad son arborícolas, que ocupan y habitan entre las superficies horizontales y verticales formados por la continuidad de los árboles o sucesión o entrecruzamiento de sus ramas, con los bejucos, epífitos y hemiepífitos, que en ritmos de periodicidad o fenología, le ofrecen los principales componentes de su dieta, refugio para el descanso e inclemencias del clima. De ese modo hacen posible la reproducción y perpetuidad de las especies tanto de primates como de los componentes vegetales. En esta delimitación de las manifestaciones de vida, entre vegetales y los primates, advertimos la estrecha relación y dependencia para la vida del bosque.

Desde el punto de vista de la concentración de agua en el suelo, en el neotrópico podemos distinguir dos grandes ecosistemas de bosques: a) bosques colinosos y no inundables, y b) bosques de bajiales o inundables. Asimismo, otros dos sistemas por la influencia de la humedad atmosférica.

En el Perú, las poblaciones naturales de primates no humanos, se hallan principalmente en la región oriental denominada Selva, que comprende el 60% del territorio nacional con casi 75 millones de hectáreas. En esta vasta región, en altitudes por debajo de 400 m.s.n.m. o selva baja, habitan 28 especies y solamente *Lagothrix flavicauda* en posible simpatía con *Aotus miconax* y *Callicebus oenanthe* (Hershkovitz 1988), habita los bosques de neblinas a altitudes entre 800 a 1100 m.s.n.m. de la parte centro-oriental. Una especie, *Alouatta pallita* habita en el bosque Occidental del Pacífico al norte del departamento de Tumbes compartiendo territorio con el Ecuador.

Las experiencias de cosechas sostenidas en el Perú desde 1976, fueron principalmente en la llanura adyacente a los grandes tributarios del río Amazonas en la parte nor-oriental, así como en el curso superior de los ríos Acre y Tahuamanú en la parte sur-oriental. Los bosques podemos diferenciarlas en dos grandes grupos como hábitat preferencial de las especies aún presentes en el Perú, es decir bosques de bajiales o inundables a los flujos estacionales de agua de los ríos y bosques de colinas o no inundables. El relieve geográfico y la composición florística son caracteres más importantes.

1. Bosques de Colina:

Este hábitat está caracterizado por la fisiografía colinosa, disecada por quebradas, con pendientes variadas que facilitan la escorrentía y el drenaje del agua de lluvia, originando los ríos con agua blanca en estación lluviosa, y de agua clara durante el estiaje. Entre la fisiografía están presentes terrazas y planicies altas o bajas, donde diferenciamos formaciones y comunidades vegetales, como los palmerales de *Mauritia* y *Phytelephas*, "varillales", "chamisales", y dominantes de *árboles*, y bosques en galerías. Especies como *Callicebus torquatus*, solamente habitan en los "varillales", cuyos componentes botánicos presentan heliofilia y esclerofilia, como resultado de su adaptación a los suelos de arena blanca. Estos "varillales" están entre los ríos Nanay e Itaya, actualmente amenazada por la agricultura desde Iquitos. También *Callicebus torquatus* habita asociada con *Pithecia* en los bosques con aspecto de varillal, de la margen izquierda en el curso medio del río Napo.

Los bosques y los palmerales entre las colinas de los ríos Manití, Tamshiyacu, Tahuayo, Tapiche, son habitados por *Cacajao calvus*, *Ateles*, *Cebus*; *Saguinus* y otros, mientras que en los bosques de las galerías en el curso superior están habitados por *Callicebus*, *Aotus* y *Saguinus fuscicollis*, *Saguinus mystax*, *Cebuella pygmaea*. Los reportes sobre *Saguinus mystax* de Soini & Soini (1982) y Glander (1984) para dos puntos diferentes del río Manití ofrecen densidades de 25-32 ind./km^2 y 9,7-29,5 ind./km^2 respectivamente, mientras que Ramírez (1984) para el río Yarapa ofrece 23,7-24,7 ind./km^2.

En el sur-oriente peruano, en la margen izquierda del río Tahuamanú, (bosque sub-húmedo tropical) se ha registrado entre 26-36 ind./km^2 de *Saguinus labiatus*.

2. Los Bosques de bajiales:

Estas formaciones ocupan terrenos adyacentes y continuos a las masas de agua lénticas y lóticas, sean ellas del tipo blanca o negra (Soini 1968), de modo que se han adaptado a la influencia de las variaciones estacionales del nivel de agua que en ocasiones alcanzan más de 4 metros. Una apreciación de las sub-cuencas con bosques expuestos a los flujos de agua negra y agua blanca está contenida en Encarnación (1985) para la llanura amazónica. Nuestras experiencias en bosques de los ríos Nanay, Napo y Tahuayo (Aquino & Encarnación 1988) nos dan densidades de 24,2 ind./km^2 y 33,0 ind./km^2 para *Aotus nancymai* y *Aotus vociferans* respectivamente. Igualmente las mayores densidades para *Saimiri boliviensis* y *S. sciureus* están dadas para los bosques inundables.

La presencia de 13 especies de primates en la Reserva Nacional Pacaya-Samiria (Soini 1986) que abarca mayor extensión de bosque inundable, nos demuestra la importancia de este tipo de bosque como hábitat para los primates.

3. Los bosques: Enfoque forestal y agropecuario:

La presencia del ecosistema boscoso está amenazada en el mundo. Los enfoques para su aprovechamiento constituyen la colonización o establecimiento de nuevos asentamientos humanos, como escape para evitar los problemas sociales de las urbes y la falta de tierras agrícolas en determinadas regiones del país.

En el Perú como en Bolivia, Brasil, Colombia y Ecuador que abarcan la cuenca amazónica, el índice de ocupación de la tierra o grado de poblamiento es bajo. Razones que justifican a los Estados para promover asentamientos humanos, cuyas metas están orientadas a la extracción de productos vegetales (frutos, plantas ornamentales y medicinales); tala selectiva de árboles maderables, construcciones de carreteras y represas hidroeléctricas, deforestación con fines agriculturales de monocultivos de arroz, café, coca, maíz, palma aceitera, implantación de ganadería, etc.

Las modalidades del uso de las tierras para cultivos diversos entre ellos pasturas, manejo forestal y protección (Cuadro N° 1) demandan la aplicación de técnicas y principios ecológicos adecuados, sin descuidar la búsqueda de beneficios para mejor la condición de vida de la población humana.

CUADRO N° 1: Capacidad de uso mayor de las tierras por Regiones en el Perú (Has/porcentaje).

	CULTIVO EN LIMPIO	PERMANENTE	PASTOS	FORESTALES	PROTECCION	TOTAL
COSTA	1140000 8,360	4960000 3,64	1622000 11,90	172000 1,26	10207000 74,84	13637000 100,0
SIERRA	1341000 3,42	20000 0,05	10576000 26,98	2092000 5,34	25169000 64,21	39198000 100,0
SELVA	2421000 3,21	219100 2,89	5718000 7,55	46432000 61,35	18924000 25,0	75686000 100,0
						128521000 100,0

El Cuadro N° 1, nos muestra que en la selva del Perú, solamente casi 19 millones de Has. estarían asignados como bosques de protección. Por los caracteres que reúnen estos bosques, están ubicados principalmente en el lado occidental de la Amazonia denominada Ceja de Selva, donde la diversidad específica de primates, así como la densidad son menores y bajas. Por similitud, los 46 millones de tierras con aptitud forestal están ubicadas en áreas de Selva Baja y Ceja de Selva, cuya fisiografía no favorece la agricultura. Sin embargo, la extracción selectiva de la madera en tiempos y períodos irregulares, la falta de asistencia y orientación técnicas en agricultura hacia los colonos, entre otros factores, van eliminando, progresiva y gradualmente, los hábitats así como los individuos de especies de primates tamaño por la acción de la cacería de subsistencia. El estado de la selva asignada al uso de cultivo en limpio que incluye las pasturas, constituyen factores que aceleran la eliminación de los hábitats así como los individuos de primates por la demanda de carnes en los asentamientos humanos. Estos ejemplos están presentes en la cuenca del Alto Mayo y Huallaga central y se acentúan al este de la cuenca del Madre de Dios. El uso de las tierras para fines de cultivos permanentes asignada solamente en unos dos millones de Has. que incluye a los cultivos de frutales como plátanos, cacao, café, cítricos y otros, están asentados en la Ceja de Selva principalmente, mientras que la palmera aceitera se promueve en la selva baja, cuenca del río Manití. De estos cultivos, los frutales comestibles por el hombre directamente al madurar, favorecen y/o enriquecen los hábitats de los primates. Pero si consideramos la mayor diversidad y población de los primates en la selva baja, es necesario la adopción y promoción de sistemas agroforestales y agrosilvopastoriles en esta llanura amazónica, a fin de obtener un adecuado uso de las tierras, elevando o mejorando su productividad, para asegurar la vida de los componentes de la fauna silvestre y diversidad biológica.

4. Las tierras de uso marginal:

En los 75 millones de Has. de tierras en la Selva, se incluye casi 3,5 millones Has. -(Perú Forestal 1987)- de terrenos pantanosos y otros menores de arena blanca a grisáceo, ubicadas principalmente en la llanura amazónica, entre los bosques de bajiales y de colina. La capacidad agropecuaria de estos terrenos es de nula a ínfima con elevada inversión en los arenales. Por la cubierta vegetal arbórea, están considerados entre las tierras de aptitud forestal; sin embargo, la ejecución de reforestación en los pantanos aún es sólo una expresión por la falta de técnicas y métodos adecuados. De modo que las cosechas sostenidas, en términos de manejo, de especies como *Cebus*, *Saimiri*, *Alouatta*, *Lagothrix*, y otras especies de la fauna silvestre, es la alternativa adecuada que justifica la conservación en términos de rentabilidad. Las comparaciones efectuadas por Vásquez (manuscrito 1990) en una hectárea de bosque de varillal cerca de Iquitos, mediante la cosecha de frutos comestibles, plantas y cortezas medicinales, plantas ornamentales e industriales para cesterka, y otros, sin disturbar la naturaleza del bosque, la rentabilidad ha significado dos veces mayor que una hectárea de terreno recién talado con cultivos de plátanos o yucas.

Estas informaciones nos llevan a una reflexión del valor e importancia que adquieren las cosechas sostenidas de animales y vegetales en un bosque marginal frente a las expectativas de la implantación de extensas áreas con monoculturas o extracción de árboles de valor como maderas.

5. El Estado de las poblaciones de primates:

Transcurridos 18 años desde Noviembre 1972, ofrecemos un avance sobre el conocimiento de las especies de primates no humanos del Perú.

En 1975 se reportada solamente 19 especies (Castro 1978), muchas poco conocidas para la comunidad científica nacional, mientras que actualmente podemos reconocer 32 especies correspondientes a 12 géneros. El conocimiento de algunas subespecies siguiendo los estudios de O. Hill (1957), P. Herskovitz (1984, 1987a,b, 1988), entre otros, es aún referencial; requiriéndose mayores esfuerzos para las exploraciones a fin de determinar la distribución geográfica y los rangos críticos en la taxonomía de las especies.

Los hallazgos que destacan son la presencia de *Saguinus labiatus* en el sur-oriente del país (Encarnación & Castro 1978) entre el curso medio de los ríos Acre-Tahuamanú, frontera con Bolivia y Brasil; *Saguinus fuscicollis* en el río Tapiche (Soini 1979), tributario derecho del río Amazonas, antes reportaba solamente para la cuenca de los ríos Yurua y Yavarí en territorio del Brasil, y las revisiones taxonómicas sobre *Aotus*, *Saimiri*, *Cacajao*, *Pithecia* y *Callicebus* de Hershkovitz (1983, 1984, 1987a, b, 1988) que ha contribuido al esclarecimiento del estado taxonómico de esas especies. De los *Aotus*, se descarta la presencia de *A. trivirgatus* y se conocen *A. nancymai* (nueva especie desde 1983), *A. vociferans*, *A. nigriceps* y *A. azarae*; pero se tiene información y registros de *A. miconax* que posiblemente presenta endemismo. De los *Saimiri*, está confirmada la presencia de dos especies con tres subespecies, *S. sciureus macrodon*, *S. boliviensis peruviensis* (nueva subespecie desde 1984) y *S. boliviensis*.

La ejecución de estudios en la Reserva Nacional Pacaya-Samiria (Soini 1982) ha comprobado la ausencia de *Saguinus mystax* en estos bosques inundables, quedando restringida para la margen derecha del río Amazonas.

En el Cuadro Nº 2, se presenta el estado actual de las especies en su medio natural, establecido en base al registro y localización, ocupación del territorio y su distribución, tamaño grupal, niveles de alopatría y simpatía, así como a los factores de tala masiva de áreas, tala selectiva, establecimiento de colonizaciones y otros, que inciden en la cacería para suministro de fuentes de alimentos. Siguiendo esos criterios y según la legislación vigente, seis especies están en situación amenazada o en vías de extinción, tres en situación vulnerable, una en situación rara; 10 en situación indeterminada y 12 en situación común o fuera de peligro.

Algunos géneros con especies cuyos individuos alcanzan pesos menores a un kilogramo, tienen amplia distribución geográfica y densidades relativas altas en sus hábitat, por consiguiente son consideradas fuera de peligro. Los géneros con especies cuyos individuos tienen más de un kilogramo de peso como *Alouatta*, *Ateles*, *Cacajao*, *Cebus* y *Lagothrix* se encuentran en situación vulnerable o amenazadas, principalmente por la presión de la cacería.

III. AREAS DE ESTUDIO Y METODOS

Desde 1973, para el reconocimiento de las especies en sus hábitat y para estimar las densidades, fue ensayado la aplicación de censos en diferentes cuencas de la Amazonia peruana. A partir de 1976, los censos fueron intensificados para seis especies de primates, dando inicio a los estudios ecológicos para desarrollar técnicas de manejo de los géneros *Saguinus* y *Cebuella* en la cuenca de los ríos Yanayacu, Tamshiyacu, Tahuayo, Yarapa, Carahuayte, Tapiche y otras. Así mismo se inicia la liberación y el manejo experimental de las poblaciones de *Saguinus mystax* en Padre Isla.

Censos poblacionales de *Aotus nancymai* fueron conducidos periódicamente de 1981 a 1987 en bosques de los ríos Nanay, Napo y Tahuayo.

Estudios sobre la cosecha sostenida de *Saguinus mystax* fueron ejecutados en 4 km^2 en el río Manití, y 1,9 km^2 en el río Yarapa, ambos afluentes derecho del río Amazonas; y 8,3 km^2 en Padre Isla entre el río Amazonas, frente a la ciudad de Iquitos. De *Aotus nancymai* el área fue en 2 km^2 ubicado en el río Tahuayo, afluente derecho del río Amazonas (Fig. 1).

Las áreas de estudio presentan bosques de bajiales, aquellos situados a orillas del río, así como bosques de altura (tierra firme) (Encarnación 1985). En todos predominan los bosques primarios, los cuales están interrumpidos por áreas de agricultura migratoria, bosques secundarios denominadas "purmas" y áreas afectadas por la extracción de madera. En Padre Isla el bosque es totalmente de bajial, inundado por el río Amazonas en época de creciente (Marzo-Junio). En las áreas de estudio y zonas de influencia están comprendidas los asentamientos humanos que son de origen mestizo, cuyas actividades económicas son la agricultura, pesca, caza, extracción de madera y colecta de frutos silvestres.

CUADRO Nº 2. Situación de las especies/sub-especies de primates no humanos en el Perú.

ESPECIE/SUB ESPECIE	NOMBRE COMUN	DISTRIBUCION REFERENCIAL	EST. DE CONSERV.*	FUENTE
FAMILIA CALLITRICHIDAE				
1. *Cebuella pygmaea*	Leoncito	Toda la Amazonia	C	Soini 1983a
2. *Saguinus fuscicollis* - *S. f. fuscicollis* - *S. f. illigeri* - *S. f. lagonotus* - *S. f. leucogenys* - *S. f. nigrifrons* - *S. f. weddelli*	Pichico	Toda la Amazonia	C	Soini 1983b
3. *Saguinus tripartitus*	Pichico	Amazonia Norte	I	Hershkovitz 1977; Thorington 1988
4. *Saguinus nigricollis* - *S. n. nigricollis* - *S. n. graellsi*	Pichico negro	Amazonia Norte	C	Hershkovitz 1977, 1982
5. *Saguinus mystax* - *S. m. mystax*	Pichico barba blanca	Amazonia Norte	C	Soini 1982
6. *Saguinus labiatus* - *S. l. labiatus*	Pichico pecho	Amazonia Sur	I	Encaranción y Castro 1978
7. *Saguinus imperator* - *S. i. subgrisescens* - *S. i. imperator*	Pichico emperador	Amazonia Sur	I	Hershkovitz 1977, 1979; Encarnación *et.al.* 1981
FAMILIA CALLIMICONIDAE				
8. *Callimico goeldii*	Supay pichico	Amazonia Central	R	Encarnación y Castro 1979; Aquino 1986
FAMILIA CEBIDAE				
9. *Aotus nancymai*	Musmuqui	Amazonia Norte	C	Hershkovitz 1983; Aquino y Encarnación 1988
10.*Aotus miconax*	Musmuqui	Amazonia Nor-Occidental	I	Hershkovitz 1983

* Codes de Estimate de Conservacion: A = amenazada, C = común, I = indeterminada, R = rara, V = vulnerable

CUADRO Nº 2. (Continuado) Situación de las especies/sub-especies de primates no humanos en el Perú.

ESPECIE/SUB ESPECIE	NOMBRE COMUN	DISTRIBUCION REFERENCIAL	EST. DE CONSERV.*	FUENTE
11.*Aotus vociferans*	Buri-buri	Amazonia Norte	C	Hershkovitz 1983 Aquino y Encarnación 1988
12.*Aotus nigriceps*	Musmuqui	Amazonia Sur	I	Hershkovitz 1983; Wright 1978, 1985
13.*Aotus azarae*	Musmuqui	Amazonia Sur	I	Hershkovitz 1983
14.*Saimiri boliviensis* - *S. b. peruviensis* - *S. b. boliviensis*	Fraile (arco romano)	Amazonia Norte, Centro y Sur	C	Hershkovitz 1984
15.*Saimiri sciureus* - *S. s. macrodon* - *C. c. discolor*	Fraile (arco gótico)	Amazonia Norte	C	Hershkovitz 1984
16.*Callicebus cupreus* - *C. c. discolor*	Tocón colorado	Amazonia Norte	C	Hershkovitz 1988
17.*Callicebus caligatus*	Tocón colorado	Amazonia Central	I	Hershkovitz 1988
18.*Callicebus brunneus*	Tocón colorado	Amazonia Sur	C	Hershkovitz 1988
19.*Cellicebus oenanthe*	Tocón colorado	Amazonia Central y Occidental	I	Hershkovitz 1988
20.*Callicebus torquatus* - *C. t. torquatus*	Tocón negro	Amazonia Norte	I	Hershkovitz 1988; Kinzey 1982
21.*Pithecia aequatorialis*	Huapo negro	Amazonia Norte	I	Hershkovitz 1987
22.*Pithecia monachus*	Huapo negro	Amazonia Central y Sur	I	Hershkovitz 1987
23.*Pithecia irrorata*	Huapo negro	Amazonia Sur	I	Hershkovitz 1987
24.*Cacajao calvus* - *C. c. ucayallí*	Huapo rojo	Amazonia Norte	A	Hershkovitz 1987
25.*Cebus apella* - *C. a. maranonis* - *C. a. peruanus* - *C. a. macrocephalus* - *C. a. pallidus*	Machín negro	Toda la Amazonia	V	Hershkovitz (en prensa)

* Codes de Estimate de Conservacion: A = amenazada, C = común, I = indeterminada, R = rara, V = vulnerable

CUADRO Nº 2. (Continuado) Situación de las especies/sub-especies de primates no humanos en el Perú.

ESPECIE/SUB ESPECIE	NOMBRE COMUN	DISTRIBUCION REFERENCIAL	EST. DE CONSERV.*	FUENTE
26. *Cebus albifrons* - *C. a. cuscinus* - *C. a. unicolor* - *C. a. yuracus*	Machín blanco	Toda la Amazonia	V	Hershkovitz 1949; Terborght, 1981
27. *Alouatta seniculus* - *A. s. seniculus*	Mono coto	Toda la Amazonia	C	Soini 1982
28. *Alouatta palliata*	Mono coto de Tumbes	Dpto. de Tumbes	A	Pulido
29. *Ateles belzebuth* *A. b. belzebuth*	Maquisapa cenizo	Amazonia Norte y Central	A	Kellog y Goldman 1944; Hershkovitz 1972
30. *Ateles paniscus* *A. p. chamek*	Maquisapa negro	Toda la Amazonia	A	Kellog y Goldman 1944; White 1986
31. *Lagothrix flavicauda*	Choro cola amarilla	Amazonia Centro y Occidental	A	Graves y O'Neill 1980; Leo 1980; Mittermeier *et.al.* 1975
32. *Lagothrix lagotricha* - *L. l. cana* - *L. l. lagotricha* - *L. l. poeppiggi*	Choro	Toda la Amazonia	A	Fooden 1963; Freese 1975; Durham 1972

* Codes de Estimate de Conservacion: A = amenazada, C = común, I = indeterminada, R = rara, V = vulnerable

Para los censos poblacionales fue aplicado el método de transectos (Freese et al. 1982). A pie, hacia el interior del bosque y en embarcaciones fluviales (Canoas) cuando los censos abarcaron las orillas de los ríos.

Para los estudios de cosechas sostenida de *Saguinus mystax* aplicamos los métodos de censo por transecto (Observador-animal)); de "chapaneo" y captura de Callitrichidae (Encarnación et al. 1978) y el de localización de los grupos en áreas predefinidas (método de mapas). Para *Aotus nancymai* se aplicó los métodos de transectos en las noches y el de captura por el aislamiento de los grupos.

Saguinus mystax alcanza su distribución en las áreas comprendidas en la margen derecha del río Ucayali y río Amazonas, y la margen izquierda del río Yavarí; el límite Sur se dispone registro para el lugar denominado "Lagarto" en la margen derecha del alto río Ucayali en 10º4'S y 73º48'O (Hershkovitz 1977).

Aotus nancymai ocurre a la margen derecha de los ríos Marañón y Amazonas y ambas márgenes del río Ucayali hasta aproximadamente 07°00'S; también alcanza al Norte del río Marañón en los "enclaves" entre parte baja y media de los ríos Tigre y Pastaza.

IV. LOS BENEFICIOS DE LAS COSECHAS SOSTENIDAS

Desde 1976, mediante las acciones coordinadas de los sectores oficiales del gobierno del Perú y de la Universidad Nacional Mayor de San Marcos, con el apoyo y la Cooperación Técnica de la Organización Panamericana de la Salud/Organización Mundial de la Salud, se está realizando las cosechas de seis especies: *Aotus nancymai*, *A. vociferans*, *Saguinus labiatus*, *S. mystax*, *Saimiri boliviensis* y *S. sciureus*. Seguidamente se presenta un ensayo de los resultados obtenidos con *Aotus nancymai* y *Saguinus mystax*.

1. Las cosechas sostenidas:

Saguinus mystax.

Especie pionera en el manejo de los primates en las áreas naturales. Las evaluaciones de los efectos de las capturas y recuperación poblacional fue en los bosques adyacentes a los ríos Manití y Yarapa.

Glander et. (1984) en la primera área reportaron que las cosechas tuvieron efectos mínimos sobre las poblaciones, cuya recuperación hacía los niveles de nueva pre-cosecha, es decir 5,0 grup./km^2 o 29,5 ind./km^2 fue en cinco años posteriores de la cosecha inicial (Cuadro N° 3).

CUADRO N° 3. Comparación de las densidades poblacionales de *Saguinus mystax* y *S. fuscicollis* en el área de estudio fundo Los Angeles - Río Manití. Período 1976 - 1982.

AÑO	S. mystax		S. fuscicollis		REFERENCIA
	Grup./Km2	Ind./Km2	Grup./km2	Ind./km2	
1976 - 77	5,0	25,0	4,7	28,0	Soini 1981(a)
1982	5,3	31,3	3,4	24,1	Glander et al. 1982

En bosques del río Yarapa se han obtenido resultados similares. La densidad pre-captura fue de 6,3/km^2 o 32,1 ind./km^2 la que fue sometida a cosecha del 66% en 1978. Los individuos remanentes en numero de 21, se reagruparon en tres grupos, de modo que en 1981, la población mostró una activa recuperación en 124% referida a la población remanente de 1978, alcanzando una densidad de 24,7 ind./km^2. En 1987, seis años después, la densidad poblacional en el área superó en un 25% a la población inicial de 1978, indicando así una completa recuperación. Ese mismo año se ha procedido a realizar una segunda cosecha del 63%. En 1989 la evaluación reportó la recuperación progresiva del 25% (Cuadro N° 4).

CUADRO N° 4 Datos de los censos sobre la recuperación poblacional de *Saguinus mystax* y *Saguinus fuscicollis* en el área de estudio del río Yarapa desde 1978 a 1989.

A Ñ O	S. mystax		S. fuscicollis		REFERENCIA
	POBLACION TOTAL	CAMBIO POBLACIONAL %	POBLACION TOTAL	CAMBIO POBLACIONAL %	
1978(1)	61	-	63	-	Moya et al 1979
1978(2)	21	-66	57	-9,5	
1981	47	124	50	-12	Ramírez 1984
1982	45	-4	59	18	
1987(1)	76	69	73	24	Moya y Puertas 1988
1987(2)	28	-63	73	24	
1989	35	25	46	-37	Puertas 1990

(1) Antes de la captura (2) Después de la captura

El análisis de la estructura poblacional de 10 grupos completos capturados en 1987 registro la composición de adultos 64,7% sub-adultos 7,5%, juveniles 22,6% e infantes 5,7%. Esta, es muy similar a la estructura de poblaciones naturales registrada para Ríos Manití, Tapiche y Tahuayo (Soini 1982). Así mismo la proporción de adultos-inmaduros fue de 1,0:0,56. Esta relación indica una tendencia a estabilizarse como una población madura tal como se reportan para las poblaciones naturales en los Ríos Manití y Tapiche de 1,0:0,5 y la del Río Tahuayo de 1,0:0,3 (Soini 1982). Estos datos nos sugiere una población vigorosa, con sucesos reproductivos ente los 8 y 9 meses. La proporción machos/hembras del nivel de adultos fue ligeramente favorable para los machos en 1,0:0,89.

Este resultado es similar al de 1981 (Ramírez 1984), pero difiere al de 1978 con mayor proporción de hembras (16%). Nuevos resultados obtenidos en 1981 y en 1987 son similares a los reportados para los Ríos Tapiche, Manití y Tahuayo (Soini 1982), lo cual explicaría el fenómeno demográfico en poblaciones adultas afectadas por las cosechas donde la recuperación de las hembras es progresiva hasta superar ligeramente a los machos, que se considera una población natural estable.

Ramírez (1984) estimó el incremento máximo poblacional de *S. mystax* al tercer año de haberse realizado la primera cosecha. Este modelo requiere un ajuste, ya que la capacidad de carga del área de estudio en 1987 alcanzó a 76 individuos, mientras que en 1978 fue de 61 individuos. Que podría significar que la tasa de incremento máximo (rm.) sea menor. Por lo tanto se requerirá mayor tiempo para la cosecha máxima; sin embargo, estimamos que las máximas cosechas se deberán realizar a partir de los 4 o 5 años después de la cosecha inicial.

La experiencia de la introducción de *S. mystax* en Padre Isla se originó como alternativa de crianza en condiciones naturales pero limitadas por una barrera física como el Río Amazonas. La población inicial liberada hasta 1980 fue de 75 individuos. En sus inicios, la adaptación y crecimiento poblacional fue lento, posteriormente este incremento fue mayor, alcanzando una población total de 161 animales, lo que significa un crecimiento promedio de 12,8% en 9 años (Cuadro N° 5).

CUADRO N° 5. Población inicial y crecimiento anual de la población de *Saguinus mystax* en Padre Isla, Loreto, Perú durante 1980 a 1989.

A Ñ O	NUMERO DE GRUPOS	TAMAÑO DE GRUPO	NUMERO DE INDIVIDUOS	COMPOSICION		INCREMENTO TOTAL (%)	CRIAS %
				A-SA-J	C		
1980	18	4,2	75	75	-	-	-
1981	14	5,9	82	77	05	9,3	6,1
1982	17	4,9	84	82	02	2,4	2,4
1983	16	5,7	91	77	14	8,3	15,4
1984	-	-	-	-	-	-	-
1985	15	6,5	98	98	-	-	-
1986	15	8,2	123	109	14	25,0	11,4
1987	15	8,4	126	114	12	2,4	9,5
1988(a)	17	9,0	153	134	19	21,0	12,4
1988(b)	17	7,6	129	122	07	-15,7	
1989	19	8,5	161	138	23	24,8	14,3

En 1988 se ejecutó la cosecha selectiva de 32 individuos. Esta primera cosecha ha sido exitosa y ha impactado positivamente en la reproducción de la especie (Cuadro N° 5). Las acciones en la protección de los bosques de la isla y de enriquecimiento con plantas de importancia alimenticia para los *S. mystax*, permiten que los niveles reproductivos de la especie se incrementen.

Para el manejo de las poblaciones naturales de *Saguinus mystax* se han considerado las secuencias:

a) La ejecución de censos pre-captura en diferentes cuencas de ríos. Estos reportaron una densidad poblacional promedio de 30 ind./km^2 o 06 grupos/km^2. (Cuadro N° 6).

b El desarrollo de una técnica de captura mediante la trampa "saguinus" de uso manual y selectiva.

CUADRO N° 6. Densidades poblacionales de *Saguinus mystax* en el Perú.

AREAS DE ESTUDIO	GRUP./KM2	IND./KM2	TRANSFERENCIAS
Río Manití	4,6	23,0	Soini 1976
Río Tahuayo	6,5	32,0	Soini 1977
Río Ucayali (J.Herrera)	6,0	36,0	Aquino 1978
Río Yanayacu	5,0	25,0	Tapia 1977
Río Carahuaite	5,3	26,5	Tapia 1978
Río Yarapa	6,3	32,1	Moya et al 1978
Río Orosa	5,7	28,5	Tapia 1979
Río Cochiquina	6,9	34,5	Moya 1981
Río Pichana	7,7	38,5	Moya 1981
Río Manití	5,3	31,5	Glander et al 1982
Río Manitk	5,1	29,9	Puertas 1988
	& = 64,4 X = 5,9	337,3 30,7	

Fluctuación 23,0 - 38,5 ind./km^2

c) La cosecha del numero de animales fue determinada en base a la densidad poblacional, afectando el 50 a 70% de la población o sea 3 a 4 grupos, quedando un remanente del 30 a 50% de la población o sea 2 ó 3 grupos, para facilitar la recuperación poblacional. En lo posible fueron capturados grupos completos, para evitar la interferencia en la organización social de los grupos.

d) La rotación de áreas de cosecha por cuencas en períodos de 4 a 5 años o más. Como ejemplos, tenemos al río Manití donde se ejecutaron cosechas en 1976, 1980 y 1988; al río Yarapa en 1978 y en 1987.

e) Alternancia de áreas de cosecha en la misma cuenca de río, dejando áreas intocadas. Una área de cosecha comprendió un tramo de 10-15 km de río y de 1 a 5 km al interior del bosque; en una o en las dos márgenes del río. En la cuenca de un río, generalmente se establecieron de 2 a 3 áreas de cosecha.

Aotus nancymai.

Las evaluaciones para los efectos de la cosecha se inició en 1981 estimando una densidad previa a la captura de 9,5 grup./km^2 o 35,5 ind./km^2. Ese mismo año fue atrapado el 59% de la población, los 29 individuos remanentes formaron 7 grupos entre completos e incompletos.

En 1984 el repoblamiento alcanzó el 169%, es decir de 49 individuos, superando a la densidad de la primera pre-cosecha en 3,5 individuos más por km^2. Ese mismo año se volvió a cosechar el 84% de la población, quedando solamente 12 individuos.

En 1987 la repoblación alcanzó el 433%, sin embargo la densidad fue inferior en 3,5 ind./km^2 menos que en 1981 y 7,0 ind./km^2 menos que en 1984. Este sugirió que la población no había logrado estabilizarse. La cosecha casi total de la población no parece haber tenido efectos negativos sobre la capacidad de repoblamiento.

Los mecanismos de repoblamiento en 1981 se han debido a la natalidad y a la inmigración de los grupos vecinos. La presencia de individuos inmaduros en por lo menos cinco grupos remanentes son evidencias de la natalidad.

En el proceso de repoblamiento en la segunda cosecha de 1984, la contribución más importante evidentemente fue la inmigración de los grupos vecinos, mientras que la natalidad jugó un rol secundario.

2. Beneficios para el bosque:

Una de las estrategias para la protección de las poblaciones de primates en el Perú, consiste en el establecimiento de unidades de conservación o parques nacionales.

En la actualidad existen cinco unidades protegidas con poblaciones de primates. El Parque Nacional Manú situada en el Sur-Oriente del País, comprende 1'532,000 Ha. y protege 13 especies de primates; el Parque Nacional Tingo María, situada en la parte central y occidental de la Amazonia, con 18,000 Hás., protege 5 especies de primates; el Parque Nacional Río Abiseo, ubicada en la parte central y occidental de la Amazonia, con una superficie de 274,520 Hás., protege principalmente a *Lagothorix flavicauda*, especie endémica que habita los bosques de neblina en la ceja de Selva; el Parque Nacional Yanachaga Chemillén, localizada en la Selva Central, con 122,000 Hás. protege 7 especies de primates; la Reserva Nacional Pacaya-Samiria, ubicada en la mitad septentrional de la Amazonia en la Selva Baja, con 2'080,000 Hás. protege 13 especies de primates (Cuadro Nº 7).

Estas áreas protegidas abarcan una superficie de 3'026,520 Hás., equivalente al 4% de la región Amazónica, protege a 17 especies. Si consideramos la diversidad total, aproximadamente el 50% de las sub-especies reconocidas están protegidas, quedando sin incluirse las otras 15 especies (Cuadro Nº 8) que justifican la creación de nuevas Unidades de Conservación.

Actualmente está en gestión la creación de la Reserva Comunal Tamshiyacu-Tahuayo, situado al Sur Este del departamento de Loreto, con una superficie de 322,500 Hás. En ésta se pretende proteger 13 especies de primates, entre ellas *Saguinus mystax* y *Cacajao calvus*, aún no representadas en las áreas protegidas por el Estado. Así mismo se viene promoviendo la creación del Parque Nacional de Loreto en la frontera con Colombia, para proteger especies cuya distribución alcanza al Nor-Este del país, y el establecimiento de una Reserva Comunal en la cuenca de los ríos Acre-Tahuamanú, Madre de Dios, para la protección de 12 especies de primates, entre ellas *Saguinus labiatus* y *Callimico goeldii*.

En el sistema de áreas protegidas, mediante la aplicación de técnicas de censos y las evaluaciones, atrapamientos, transporte y otros, según las necesidades de uso de las especies de primates, son aplicables las cosechas sostenidas, las que afianzarán la razón de dichas áreas orientado al mayor rendimiento del bosque.

CUADRO Nº 7. Especies de primates no humanos protegidos en las Unidades de Conservación del Perú.

ESPECIE	UNIDADES DE CONSERVACION				
	P.N. Manu	P.N. Tingo Maria	P.N. Rio Abiseo	P.N. Yanachaga Chemillen	P.N. Pacaya Samiria
Cebuella pygmaea	X	X			X
Saguinus fuscicollis	X	X		X	X
Sagunius imperator	X				
Callimico goeldii	X				X*
Aotus nancymai					X
Aotus nigriceps	X			X	
Saimiri boliviensis	X	X		X	X
Saimiri sciureus					X*
Callicebus brunneus	X				
Pithecia monachus	X				X
Cebus apella	X	X		X	X
Cebus albifrons	X		X		X
Alouatta seniculus	X	X		X	X
Ateles belzbuth			X		X
Altes paniscus	X			X	X
Lagothris flavicuda			X		
Lagothris lagothricha	X			X	X

* Aún no comprobadas in situ.

También la cosecha sostenida garantiza el equilibrio de la vida en el bosque por la facilidad de recuperación de las poblaciones de las especies afectadas, teniendo en cuenta la ausencia de la cacería.

CUADRO Nº 8. Especies de primates no humanos no protegidos en las Unidades de Conservación del Perú.

Saguinus tripartitus
Saguinus nigricollis
Saguinus mystax
Saguinus labiatus
Aotus miconax
Aotus vociferans
Aotus azarae
Callicebus cupreus
Callicebus caligatus
Callicebus oenanthe
Callicebus torquatus
Pithecia aequatoriales
Pithecia irrorata
Cacajao calvus
Alouatta palliata

3. Beneficios para la humanidad:

Las cifras de exportación de primates desde Iquitos hasta 1974, sin medidas de manejo, nos demuestra una clara tendencia del beneficio directo hacia la población humana de los países desarrollados cual haya sido el uso, aún cuando las investigaciones biomédicas significan beneficio a toda la humanidad.

El inicio de la cosecha sostenida desde 1976 en el Perú, es el inicio de la experiencia para compartir los beneficios entre el poblador rural y poblador urbano. Al mismo tiempo ha promovido y facilitado el apoyo en términos de cooperación técnica desde los países desarrollados, buscando algunos patrones de equilibrio y compensación entre países productores y países usuarios de primates no humanos.

Las evidencias de este intercambio se manifiestan la incorporación e integración de los pobladores rurales en las actividades de cursos, evaluaciones, atrapamientos, retención, transporte y otros, con bases y fundamentos técnicos a cargo de especialistas en recursos naturales y manejo. También la construcción de un zoocriadero y su equipamiento para su funcionamiento, con los resultados positivos del manejo en cautiverio de las especies de interés, ofrecen oportunidades de: a) Entrenamiento y capacitación a estudiantes de todo nivel, b) Prestación de servicios de laboratorios básicos hacia la comunidad, c) Ejecución de investigaciones biológicas y eco-etológicas.

Las acciones enunciadas significan aportes en educación ambiental y ecología para los adultos en general, difíciles de ejecutar al presente en los Villorios de la Amazonia. Por otro lado, fortalece el intercambio socio-cultural entre los países del mundo.

V. CONCLUSION

En resumen se puede concluir que en una población natural de *Aotus*, la cosecha de un numero razonable de individuos, la tala y el desbroce selectivo en la captura, no ocasionan efectos negativos sobre la población. Así mismo la vegetación muestra una recuperación rápida, inclusive permitiendo el uso de las madrigueras como ambiente para el "sueño diurno".

El aprovechamiento máximo del potencial de los bosques de la Amazonia deberá realizarse manteniendo en lo posible su estado natural, sin alterarlo o destruirlo por la extracción de uno o varios de los productos, tal como ocurre con la extracción selectiva de madera o la construcción de áreas de agricultura. Las experiencias en el manejo sostenido de las poblaciones de *Saguinus mystax* y *Aotus nancymai* nos demuestran las grandes posibilidades de realizar cosechas permanentes y a largo plazo sin afectar los bosques.

VI. REFERENCIAS

Aquino, R. 1990. La Fauna Primatológica en áreas de Jenaro Herrera, páginas 122-134. In *La Primatologia en el Perú, Investigaciones Primatológicas (1973 - 1985), Proyecto Peruano de Primatología, "Manuel Moro Sommo"*, N. E. Castro-Rodríguez, ed. Lima, Perú: Dirección General de Forestal y Fauna, Ministerio de Agricultura; Instituto Veterinario de Investigaciones Tropicales y de Altura, Universidad Nacional Mayor de San Marcos; Organización Panamericana de la Salud, Organización Mundial de la Salud.

Aquino, R. y Encarnación, F. 1986. Population and structure of *Aotus nancymai* (Cebidae: Primates) in Peruvian Amazon lowland forest. *Am. J. Primatol.* **11**, 1-7.

Aquino, R. y Encarnación, F. 1988. Population densities and geographic distribution of night monkeys (*Aotus nancymai* and (*Aotus vociferans*) (Cebidae: Primates) in Northeastern Peru. *Am. J. Primatol.* **14**, 375-381.

Aquino, R, Puertas, P. and Encarnación, F. en prensa. Efectos de las cosechas sobre las poblaciones de *Aotus nancymai* (Cebidae, Primates) en bosque inundable. Iquitos, Perú: Centro de Reproduccion y Conservacion de Primaes No Humanos, Instituto Veterinario de Investigaciones Tropicales y de Alturas (IVITA)

Castro, N. E. 1978. Diagnóstico de la situación actual de los primates no humanos en el Perú y un plan para su utilización nacional, 20 pág. Lima, Perú: Ministerio de Agricultura y Alimentación

Encarnación, F., Moya, L., Soini, P., Tapia, J., and Aquino, R. 1978. Captura de Callitrichidae *Saguinus* y *Cebuella* en la Amazonía Peruana. *Boletín de Comité Regional de Coordinación del Proyecto Primates* **1**, 1-20.

Encarnación, F. 1985. Introducción a la flora y vegetación de la Amazonía Peruana: Terminología y breve descripción de las formaciones vegetales de la Llanura Amazónica. *Candollea* **40**, 237-252.

Encarnación, F. y Castro, N. 1990. Informe preliminar sobre Censo de Primates no Humanos en el Sur-Oriente Peruano: Iberia e Iñapari (Departamento de Madre de Dios), Mayo 15 - Junio 14, 1978, páginas 57-67. En *La Primatologia en el Perú, Investigaciones Primatológicas (1973 - 1985), Proyecto Peruano de Primatología, "Manuel Moro Sommo"*, N. E. Castro-Rodríguez, ed. Lima, Perú: Dirección General de Forestal y Fauna, Ministerio de Agricultura; Instituto Veterinario de Investigaciones Tropicales y de Altura, Universidad Nacional Mayor de San Marcos; Organización Panamericana de la Salud, Organización Mundial de la Salud.

Freese, C. 1975. A Census of Nonhuman Primates in Peru, páginas 17-71. En Primate-Censusing Studies in Peru and Colombia, Report to the National Academy of Sciences on the Activities of the Project, (1) AMRO-0719. Washington DC, USA: Organización Panamericana de la Salud (PAHO).

Glander, K., Tapia, J. y Fachín, A. 1984. The impact of trapping and removal of wild populations of *Saguinus mystax* and *Saguinus fuscicollis*. *Am. J. Primatol.* **7**, 89-97.

Hershkovitz, P. 1977. *Living New World Monkeys (Platyrrhini)*, vol. 1, 1117 pages. Chicago, IL, USA: University of Chicago Press.

Hershkovitz, P. 1983. Two new species of night monkeys, genus *Aotus* (Cebidae, Platyrrhini): A preliminary report on *Aotus* taxonomy. *Am. J. Primatol.* **4**, 209-213.

Hershkovitz, P. 1984. Taxonomy of squirrel monkeys, genus *Saimiri* (Cebidae, Platyrrhini): A preliminary report with description of a hitherto unnamed form. *Am. J. Primatol.* **7**, 155-210.

Hershkovitz, P. 1987a. Uacaries, New World monkey of the genus *Cacajao* (Cebidae, Platyrrhini): A preliminary taxonomic review. *Am. J. Primatol.* **12**, 1-53.

Hershkovitz, P. 1987b. The taxonomy of South American sakis, genus *Pithecia* (Cebidae, Platyrrhini): A preliminary report and critical review with the description of a new species and new subspecies. *Am. J. Primatol.* **12**, 387-468.

Hershkovitz, P., 1988. Origin, speciation, and distribution of South American titi monkeys, genus *Callicebus* (Family Cebidae, Platyrrhini). *Proc. Acad. Nat. Sci. Phil.* **140(1)**, 240-272.

Hill, W. C. O. 1960. Primates: Comparative Anatomy and Taxonomy IV., *Cebidae*, Part A, a Monograph. Edinburgh, Scotland: University Press, xxii, 523 pág.

Moya, L., Verdi, L., Bocanegra, G. y Rimachi, J. 1979. Análisis poblacionales de *Saguinus mystax* (Spix, 1823) (Callitrichidae) en la cuenca del Río Yarapa, Loreto, Perú. *Boletín de Comité Regional de Coordinación del Proyecto Primates* **2(1)**, 1-44.

Moya, L. y Puertas, P. 1988. Evaluación poblacional y cosecha experimental de *Saguinus mystax* en bosque del Río Yarapa en la Llanura Amazónica Peruana. pp. 1-13. Iquitos, Perú: Proyecto Peruano de Primatología.

Ramírez, M. 1984. Population recovery in the moustached tamarin (*Saguinus mystax*: Management strategias and mechanism of recovery. *Am. J. Primatol.* **7**, 245-259.

Soini, P. 1990. Nota sobré el hallazgo de una sub-especie adicional de *Saguinus fuscicollis* (Callitrichidae, Primates) para el Perú, páginas 314-317. En *La Primatologia en el Perú, Investigaciones Primatológicas (1973 - 1985), Proyecto Peruano de Primatología, "Manuel Moro Sommo"*, N. E. Castro-Rodríguez, ed. Lima, Perú: Dirección General de Forestal y Fauna, Ministerio de Agricultura; Instituto Veterinario de Investigaciones Tropicales y de Altura, Universidad Nacional Mayor de San Marcos; Organización Panamericana de la Salud,

Organización Mundial de la Salud.

Soini, P. y Soini, M. 1990. Distribución y ecología poblacional de *Saguinus mystax* (Primates, Callitrichidae), páginas 272-313. En *La Primatologia en el Perú, Investigaciones Primatológicas (1973 - 1985), Proyecto Peruano de Primatología, "Manuel Moro Sommo"*, N. E. Castro-Rodríguez, ed. Lima, Perú: Dirección General de Forestal y Fauna, Ministerio de Agricultura; Instituto Veterinario de Investigaciones Tropicales y de Altura, Universidad Nacional Mayor de San Marcos; Organización Panamericana de la Salud, Organización Mundial de la Salud.

Tapia, J., Encarnación, F., Aquino, R., Moya, L. y Soini., P. 1990. Censos poblacionales y sacas periódicas de primates en la Amazonía peruana (1976-1985), pp. 325-341. In *La Primatologia en el Perú, Investigaciones Primatológicas (1973 - 1985), Proyecto Peruano de Primatología, "Manuel Moro Sommo"*, N. E. Castro-Rodríguez, ed. Lima, Perú: Dirección General de Forestal y Fauna, Ministerio de Agricultura; Instituto Veterinario de Investigaciones Tropicales y de Altura, Universidad Nacional Mayor de San Marcos; Organización Panamericana de la Salud, Organización Mundial de la Salud.

FACTORES AMBIENTALES QUE AFECTAN A LA REPRODUCCIÓN DE LOS PRIMATES EN HÁBITAT ESPECIALES

Thomas L. Wolfle

Instituto de Recursos Animales para Laboratorios
Consejo Nacional de Investigación, Academia Nacional de Ciencias,
Washington, D.C.

Esta presentación se concentra en tres elementos clave de conservación y manejo efectivo y humanitario de primates no humanos: el ambiente, los hábitat para la investigación en cautiverio, y el comportamiento individual y típico de las especies. Es esencial adquirir conocimientos sobre cada uno de estos. Sin embargo, para que se puedan combinar en una estrategia eficaz, es preciso añadir a la fórmula un cuarto elemento, el hombre. El manejo de las colonias cautivas de primates no humanos requiere un conocimiento de los comportamientos típicos de cada especie, y el impacto de los diferentes ambientes de cautiverio en cuanto al desarrollo o inhibición del comportamiento. La traducción de los comportamientos típicos de las especies expresadas por los animales en su estado salvaje a los animales en cautiverio, es un punto de partida obvio aunque no sin riesgos. El entendimiento de los comportamientos en cautiverio que son diferentes de los demostrados en estado salvaje, pero no necesariamente anormal, es un campo que necesita considerable investigación.

Generalmente se considera que los comportamientos estereotípicos son indicativos de reacciones a la mala adaptación al estrés de un ambiente en cautiverio inadecuado. Ejemplos probables de ello son la automutilación y el infanticidio. Sin embargo, otros, como el pasear o dormir, pueden ser comportamientos que le permiten al animal adaptarse eficazmente al nuevo ambiente. El entendimiento de la dinámica de estas reacciones a los ambientes en cautiverio es una difícil cuestión que requiere investigaciones colaborativas entre expertos en comportamiento, primatólogos, endocrinólogos, fisiólogos en reproducción, técnicos y otros. Los administradores de colonias encargados de la atención humanitaria y la cría de primates no humanos deben enunciar y documentar claramente un programa en el que se discuta el bienestar fisiológico y de comportamiento de los animales. Este programa deberá identificar las metas, así como opciones para alcanzarlas, y delegar autoridad específica a quienes poseen las aptitudes profesionales necesarias para su interpretación y ejecución.

Introducción

Cuando Linneo dió el nombre de Primates al orden taxonómico que incluye al hombre, los monos y los simios, sin duda alguna estaba impresionado con la diversidad de este "primer" orden del reino animal, que incluye diferencias de conducta, morfológicas y sociales de enormes proporciones. Morfológicamente, el tamaño de los primates no humanos oscila desde menos de un kilo hasta más de 100; algunos tienen colas prensiles; otros se han adaptado a un estilo de vida arbóreo que requiere el uso de los brazos, mientras que otros habitan en el suelo. Desde el punto de vista social y de conducta, estos incluyen a los que son activos de noche así como a los activos de día; los que viven una vida solitaria y los que son sociales; los que se aparean de por vida y aquéllos cuya estructura social favorece el apareamiento colectivo; los que fabrican nidos y los que duermen en lugares improvisados; y aquéllos cuya estructura jerárquica la determinan líneas matriarcales o patriarcales, el tamaño y la fuerza o la posición en la familia.

Estas y otras diferencias son la base de la variedad de comportamientos típicos observados en más de cuarenta especies de primates no humanos encontrados en los ambientes de investigación en cautiverio.. Representan un enorme desafío para los administradores de las colonias, que deben construir jaulas y ambientes adecuados que le permitan a los animales desarrollar los comportamientos típicos de sus especies. Generalmente se considera que la expresión de los comportamientos típicos de las especies es una manifestación deseable de la aptitud de un animal para adaptarse eficazmente dentro de su cautiverio. Por lo tanto, el costo en energía para que un animal se adapte a un ambiente nuevo es una consideración esencial cuando se determina la adecuación del ambiente y la cantidad de estrés que éste (y el proceso de adaptación) impone sobre el animal. A fin de cuentas, la capacidad del animal para adaptarse a un ambiente nuevo con un estrés mínimo continuo es la meta deseada, tanto desde el punto de vista del bienestar del animal como de su estabilidad fisiológica y bioquímica esenciales para la investigación.

Es obvio que los animales están bajo cierto grado de estrés, ya sea en su estado salvaje o en cautiverio. Por ejemplo, los estrés conectados con las relaciones entre presa y predador y con la búsqueda de alimentos en la selva son, hasta cierto punto, reemplazados en el cautiverio con los estrés resultantes del aburrimiento, la falta de interacción social apropiada o las restricciones de espacio. La meta no es, en consecuencia, identificar y aliviar cada fuente de estrés, sino determinar los cambios de conducta, fisiológicos y bioquímicos indicativos de un proceso homeostático que no puede retornar el animal a una base de referencia estable, lo que significa que el animal no se puede adaptar.

Estos comentarios no tienen por objeto abarcar toda esta materia, sino más bien establecer una base conceptual sobre la cual formular objetivos y acciones de conformidad con las metas de esta primera reunión del Comité Regional de Primatología para las Américas. A continuación se amplifican estos conceptos introductorios y se proporciona una estrategia mediante la cual abordar más a fondo las distintas disciplinas.

Agrupación o división

A veces parece que las discusiones sobre este tema son incompatibles en el sentido que primero recalcan la importancia de identificar la variabilidad individual que existe entre los animales de una especie (división), y luego examinan especies e incluso familias u órdenes como si no hubiese diferencias importantes (agrupación). En realidad, ambos conceptos son válidos e igualmente importantes. Hay tanta variabilidad individual entre, por ejemplo, los monos ardilla (*Saimiri*) o monos nocturnos (*Aotus*), incluso dentro de aquéllos de la misma región geográfica o subespecie, que constantemente se nos recuerda la necesidad de aprender las diferencias individuales de los comportamientos de cada animal en una colonia para poder juzgar las reacciones de ese animal en un proyecto o ambiente de investigación. Estas diferencias individuales frecuentemente, pero no invariablemente, están relacionadas con experiencias derivadas del ambiente en cautiverio y se pueden manifestar como un aumento o disminución en la agresión hacia los compañeros en la jaula o hacia sí mismos, el temor y el estrés relacionados con la proximidad de personas o la restricción por razones médicas o experimentales, o el aumento o reducción en apetito o actividad. Si se conoce el punto de referencia de estos u otros comportamientos antes de colocar un animal en una jaula común con otros, o en un protocolo de investigación, se puede contar con un aviso adelantado cuando el animal manifieste niveles inadmisibles de estrés e indicar que se requiere cierta intervención. Debe recalcarse que el reconocimiento de cuándo intervenir es a menudo difícil, a menos que se conozcan estas diferencias individuales, e incluso entonces se debe contar con una determinación profesional. Los técnicos de investigación o en atención animal generalmente son los más indicados para tomar estas decisiones, porque a menudo son los más conocedores de los animales bajo su cuidado. También hay otra consideración. El alivio del estrés se puede abordar desde el punto de vista del bienestar del animal o de la confiabilidad de los datos de investigaciones. Puede ser que un animal se ocupe activamente en actividades (de conducta o fisiológicas) que le permiten hacer frente al ambiente existente, pero que son incompatibles con las investigaciones. Por consiguiente, tal vez sea preciso intervenir para aliviar el estrés y mantener la estabilidad de la investigación, aun cuando el animal aparentemente se esté adaptando bien al ambiente. Dependiendo de la naturaleza de las investigaciones, también puede suceder lo contrario -- puede que sea necesario intervenir por el bien del animal, aunque la investigación no corra ningún peligro. En cada uno de estos casos, las metas de las investigaciones y el criterio del científico, y del veterinario, así como la atención y el uso de los animales son consideraciones esenciales que deben identificarse claramente y discutirse en un plan institucional.

Habiendo dicho esto sobre las diferencias individuales, las semejanzas entre las especies son, sin embargo, dignas de consideración, pues integran el conocimiento básico que la gente tiene de los animales. Las diferencias entre los macacos y los tití se entienden principalmente debido a un conocimiento de las especies representativas, como los monos rhesus (*Macaca mulatta*) y tití de barriga roja (*Saguinus labiatus*), respectivamente, independientemente de las diferencias individuales entre los miembros de cada especie. Cuando se planea la reproducción o alojamiento de una especie de primates no humanos, los requisitos para la vivienda y el mantenimiento se basan principalmente en un conocimiento de las especies y, en menor grado, en una comprensión de la variedad de diferencias individuales que es probable encontrar en el grupo. La conveniencia de viviendas para alojar parejas, familias o grupos grandes depende principalmente de la especie, pero la necesidad de otro tipo de vivienda es una función de las diferencias individuales y los conflictos dentro del grupo. Por consiguiente, es muy obvio que antes de recibir una colonia, o un solo individuo, se debe conocer a fondo la especie de que se trate.

De esto se debe deducir claramente que no basta con "dividir" la especie y definirla únicamente como que consta de un grupo de diferencias individuales. Ni tampoco es suficiente "agrupar" a todos los individuos de una especie en un solo prototipo. Obviamente, el manejo y uso humanitario adecuado requiere un conocimiento completo de los rasgos típicos de las especies y de las diferencias individuales.

Un plan de acción - El plan institucional

Para determinar lo que se debe hacer respecto de la vivienda, atención y cría de los primates no humanos, es preciso formular un plan de acción sobre cómo proceder al efecto. Este "plan institucional" tiene varios componentes claves que deben estar en orden antes de poner en perspectiva aspectos específicos como el ambiente, el hábitat y la reproducción.

Aunque con el transcurso del tiempo cualquier plan puede abarcar mucho, distintos planes institucionales varían considerablemente en función de muchos factores. Sin embargo, un tema invariable es que no basta con concentrarse en un solo parámetro, tal como la reproducción. En este ejemplo, la reproducción es el resultado final de una colonia bien administrada que aborda el bienestar de cada individuo. El bienestar, no la reproducción, es la meta del plan. Si bien la reproducción es un indicador importante, e incluso crítico, un plan que solo considere los parámetros reproductivos y excluya otros factores ambientales, sociales, nutricionales y de crianza, es de muy corta visión. Aunque para algunos puede resultarles sorprendente, el bienestar es algo más que reproducción y carencia de dolor y malestar. Toda evaluación del bienestar de los animales dentro de un ambiente cautivo debe tener en cuenta muchos otros parámetros, siendo el más importante de ellos la motivación y el adiestramiento de quienes atienden y usan esos animales.

La estabilidad de las bases en que se sustenta la adecuación de los programas de conservación, y el manejo y la salud de los primates no humanos en cautiverio, depende de la claridad de la expresión y aplicación de un programa institucional que enuncie las metas, la responsabilidad y las líneas de autoridad. Del mismo modo que no es aceptable traer nuevas especies de primates a un ambiente cautivo sin la enseñanza y el adiestramiento adecuados de quienes cuidarán a esos animales, tampoco es aceptable iniciar o llevar a cabo medidas de conservación, cría o investigación con primates sin una política y metas claramente escritas y enunciadas. Esta declaración, hecha por el funcionario institucional principal a cargo del programa, debe identificar a una sola persona responsable de todo el programa, con autoridad para comprometer recursos esenciales de personal y económicos para alcanzar el propósito fijado. Dado que la mayoría de esos programas afectan a diferentes departamentos dentro de una universidad o empresa, la persona responsable debe ocupar un cargo dentro de la administración, con autoridad por encima de cada uno de los departamentos componentes. El plan debe contener una declaración precisa de las metas del proyecto, con puntos de referencia suficientemente detallados como para poder medir el grado de realización, o desviación, de las metas. Los títulos de capítulos en la *Guía para el cuidado y uso de animales de laboratorio* del ILAR generalmente se emplean en planes formulados por instituciones que utilizan animales en investigaciones, pruebas o capacitación. Podría elaborarse una guía similar para programas de conservación, partiendo de principios o prácticas de la conservación, como las Normas internacionales para la adquisición, cuidado y cría de primates no humanos de la Sociedad Internacional Primatológica.

El personal es la clave de los programas de conservación, cría y manejo. Aunque a menudo se citan las instalaciones como esenciales para programas "dignos de crédito", actualmente se reconoce la valía del personal sobre la de los ladrillos y el mortero de la instalación. Este énfasis en el personal se inicia con el compromiso contraído por el individuo responsable del programa y el grado al cual esa persona transmite ese compromiso a quienes están bajo su cargo. El variado personal que directa o indirectamente afecta a los animales o su ambiente es de igual importancia. Entre estas personas figuran no solo investigadores, biólogos, técnicos y veterinarios que interactúan directamente con los animales, sino también ingenieros, secretarias, guardias de las instalaciones y el personal de saneamiento que tiene contacto casual con los animales o que de una manera u otra puede tener necesidad para entrar en la instalación donde están los animales, o cuyo trabajo puede depender de esa instalación. La capacitación de este último grupo puede beneficiar directamente a los animales mediante el mantenimiento de instalaciones por los ingenieros de edificios y saneamiento, o indirectamente a través de expresiones públicas de apoyo a los programas por las secretarias y el personal auxiliar de las oficinas. El plan institucional debe identificar a cada uno de estas personas y el adiestramiento o enseñanza apropiado para cada una de ellas.

Tipos de enjaulado y propensiones sociales

La primera consideración cuando se planean viviendas para primates no humanos es si los animales se alojarán solos, en parejas o en grupos sociales. Aunque existen excepciones, la mayoría de las especies de primates no humanos viven en grupos sociales en la selva y se beneficiarán de la oportunidad de tener cierta interacción social con individuos de su misma especie. Como se indicó anteriormente, la fuerza de unión de la estructura social se puede establecer mediante líneas familiares o jerárquicamente a través de la fuerza bruta o de un macho "alfa". Pueden ser unidades estrechamente unidas o agregados sueltos, y pueden constar de una sola pareja o de una tropa de más de cien. Estas características típicas de las especies se deben conocer en relación con cada especie manejada, criada o conservada.

Una vez que se decide un sistema de vivienda, es preciso formular una estrategia para introducir en ella a los animales y observarlos. No basta con "meter" a los primates en un grupo social y esperar a que nazcan los bebés, aunque el caos intencional ocasionado por múltiples introducciones simultáneas sí parece reducir la agresión entre los miembros de la colonia en relación con la que muestran hacia los individuos introducidos individualmente. Es necesario formular un método para la adaptación entre los miembros del grupo social planificado, a fin de reducir al mínimo la agresión y el trauma de animales incompatibles. Una vez formado el grupo social, la observación frecuente debe asegurar la identificación de los animales excesivamente agresivos o traumatizados y su extracción si fuera necesario.

Independientemente de la configuración de la vivienda y el hábitat que se establecerán para una especie, existe una ecología más grande que afecta con fuerza considerable a los primates. Esta ecología consta de la habitación circundante, sus habitantes, incluido el personal, y el alumbrado, calefacción y sistema de ventilación. Los administradores de colonias de primates son bien conscientes de la influencia de un animal situado en las jaulas de enfrente. No solo se establecen jerarquías dentro de grupos de la misma especie, sino que también se establecen mediante expresiones de comportamiento, sonidos y olores con animales de jaulas cercanas. El traslado a la jaula de enfrente de un individuo excesivamente dominante de un grupo alojado por pares, servirá de poco para reducir las tensiones en el individuo sumiso. Un animal extraído momentáneamente de un grupo social para recibir tratamiento médico menor debe ser colocado lo más cerca posible del grupo para reducir el trauma asociado con el restablecimiento de la jerarquía social. De igual modo, el retorno de uno o un par de animales a las jaulas después de limpiarlas, debe tener en consideración el equilibrio social y ecológico de la ubicación de la jaula original y no reubicarlos indiscriminadamente dentro de la habitación. Es importante entender este concepto a fin de reducir el innecesario estrés de conducta o física en los animales, y mejorar el bienestar mediante la aplicación de principios ecológicos conducentes al bienestar de la especie.

El personal, en particular los técnicos conocidos por los animales, desempeña una función importante en la ecología del ambiente de la habitación. Las personas comprensivas y bien adiestradas pueden ser una fuente de consuelo y estabilidad para la habitación, y deben hacer el papel del individuo "alfa" dentro de la habitación. De igual modo, las personas desconsideradas o mal entrenadas pueden ser una fuente constante de malestar para los animales y de considerable variabilidad para los estudios a los cuales están asignados los animales.

Entre las otras tendencias sociales que requieren atención figuran las relacionadas con el apareamiento, la reproducción y el juego. Cada uno de estos comportamientos típicos de especies requiere la atención de personal conocedor para asegurar su expresión apropiada.

Reproducción

La reproducción es tanto la meta de un programa de crianza como una medida de la adecuación del hábitat y la ecología en que se mantienen los animales. La reproducción a menudo se cita como el indicador necesario del bienestar y la adecuación de la jaula o el ambiente. Es muy probable que esto no sea cierto. Las pruebas sugieren que algunos comportamientos, como el apareamiento y la reproducción, atención de lactantes, comida y eliminación son tan fuertes que se expresan en ambientes inadecuados. Las evaluaciones de estos parámetros de la conducta solo llevarían a concluir erróneamente que no es necesario mejorar o cambiar el ambiente. Muchos de esos animales probablemente están alojados de esta forma actualmente. Sin embargo, cuando se proporciona un ambiente mejor a otros animales como medio para evaluar el anterior, las diferencias es probable que se hagan obvias. Una manera de realizar esto consiste en proporcionar opciones a los animales, así como niveles de control de algunos aspectos de su ambiente, para construir un patrón de condiciones aparentemente muy apropiadas para esa especie. Este patrón, formulado pidiéndole al animal que exprese preferencias, combinado con observaciones fisiológicas, endocrinológicas y de conducta, proporcionarán una seguridad razonable de que la situación es aceptable para el animal y que su estado de bienestar es lo más próximo posible al óptimo.

El costo de la adaptación

El concepto del gasto de energía es digno de más análisis. Los mecanismos homeostáticos de los animales en todos los ambientes se esfuerzan constantemente por retornar al animal a un nivel básico de comodidad y bienestar. Cuando este proceso tiene éxito en permitir que el animal se adapte, sin costo excesivo en conducta o fisiológico, hay pocas dudas en cuanto al animal o el ambiente. Sin embargo, cuando se hace evidente que esta adaptación solo se logra a un gran costo, se justifica la intervención. La prueba de que esto ocurra no se puede obtener mediante la observación casual o no informada, ya que a menudo el animal no iza una bandera roja, al menos en las primeras etapas. El aviso mejor y más anticipado es a través de la observación astuta de cambios de conducta sutiles por personas que conozcan al animal. Estos cambios pueden incluir indicadores tales como una mayor actividad, aumentos o disminuciones en el aseo personal, los cambios en el ciclo sueño-despierto, la personalidad alterada o la agresión, así como en el comer.

Los movimientos repetitivos, a menudo llamados comportamientos estereotípicos o estereotipos, en general se consideran indicadores de que el ambiente no es el correcto para el animal. Aunque esto puede ocurrir, también debe recordarse que estos comportamientos pueden ser expresiones de adaptación al ambiente. Un ejemplo de esto a menudo se relaciona con rutinas de alimentación en cautiverio. En la selva, los primates pasan casi el 80% de su tiempo en busca de alimentos y comiendo. En cautiverio, este periodo frecuentemente se reduce a menos del 10%. ¿Qué hace el animal con el resto del tiempo? Puede pasear, descansar, asearse o dedicarse a otros comportamientos que ocupan tiempo. Como se pasan tanto tiempo en estos comportamientos, puede pensarse que estos son estereotipos producidos por una mala adaptación, cuando en realidad pueden ser estrategias muy eficaces para distribuir su nueva cuota de tiempo. Todavía se necesita estudiar más este campo antes de poder formular recomendaciones concluyentes. Mientras tanto, es preciso continuar con las cuidadosas observaciones e interpretaciones que están llevando a cabo personas conocedoras de las especies y los animales individualmente.

Medidas fisiológicas y bioquímicas

A veces es posible detectar cambios en las hormonas pituitarias y adrenales durante las etapas agudas de dolor o estrés, pero a medida que el estrés se hace crónico, los indicadores bioquímicos se hacen mucho más variables, y pueden revelar aumentos, disminuciones o ningún cambio. Las hormonas pituitarias a menudo son los indicadores más fiables del dolor agudo y el estrés, se puede esperar que incluyan una disminución en los niveles de insulina, junto con mayores niveles de prolactina, hormonas de crecimiento, glucagón y epinefrina. Esta combinación de cambios contribuye al desarrollo de hiperglucemia, que se puede utilizar en un perfil de evaluación de los animales. El sistema simpatoadrenomedular también puede responder al dolor y estrés mediante el aumento de los niveles de catecolaminas. Las consecuencias fisiológicas de la activación de estos sistemas incluyen un mayor rendimiento cardíaco y flujo sanguíneo de los músculos esqueléticos, la vasoconstricción cutánea, reducción en el movimiento intestinal y más glucosa disponible.

De este breve recuento no se debe suponer que unas pocas medidas fisiológicas y bioquímicas pueden servir de indicadores fiables del estrés o el bienestar. Solo pueden ayudar en la consideración de muchas otras observaciones de la conducta y el ambiente.

Comportamientos por mala adaptación

Llegado cierto momento, ya sea mediante la observación directa o por apreciación profesional, se puede considerar que un comportamiento es el resultado de una mala adaptación y ha dejado de serle de utilidad práctica al animal. Entre los comportamientos obvios de este tipo figuran la automutilación, el infanticido, el acicalamiento excesivo que termina en tirarse del pelo, la agresión y la sumisión excesiva. Cuando se determina que estos comportamientos no son ocasionados por dolor o una enfermedad, deben examinarse las causas ambientales y del desarrollo que los predisponen. Raras veces se pueden aliviar estas manifestaciones por medios farmacológicos, aunque los tranquilizantes algunas veces son eficaces para ayudar al animal a hacerle frente a un cambio en el ambiente, instituido para lograr una modificación en comportamiento.

El cambio en el tamaño y la complejidad de la jaula frecuentemente son las primeras consideraciones para resolver los problemas producidos por los comportamientos debidos a la mala adaptación. Los tamaños de las jaulas generalmente son seleccionados de manera que los animales puedan expresar los comportamientos típicos de su especie. Si bien esta práctica puede ser encomiable e indicativa de un alto nivel de interés en el bienestar del animal, hay pocas pruebas que sugieran que por necesidad hay que elaborar comportamientos típicos de las especies para asegurar su bienestar. Además, una vez que se establezcan comportamientos debidos a una mala adaptación, es probable que no respondan satisfactoriamente a tan solo un cambio en el tamaño de la jaula. El tamaño de la jaula, por sí solo, no parece ser el contribuyente principal a los comportamientos por mala adaptación. Esto se ha demostrado en demasiadas colonias en las cuales primates lactantes fueron criados sin padres, en jaulas suficientemente grandes para el desarrollo de comportamientos normales. En este y en muchos otros casos de comportamientos por mala adaptación, la raíz del problema radica en la falta de interacción social adecuada durante los periodos críticos del desarrollo. Por lo tanto, los problemas quizás no se deban a las condiciones actuales, sino que pueden haber surgido de deficiencias sociales en una edad mucho más temprana.

Este ejemplo proporciona un poco de información que puede utilizarse actualmente. Cuando se sabe que las condiciones de la crianza llegan a producir comportamientos por mala adaptación en un número considerable de animales en una edad posterior, estas se deben modificar de inmediato aunque la actual población de animales se muestre aparentemente normal.

Importancia de la adaptación para los datos de la investigación

Muchos animales pueden parecer que se están adaptando satisfactoriamente a la vivienda de la colonia, sin mostrar cambios de comportamiento o fisiológicos alarmantes. Sin embargo, aun estos animales están bajo estrés, cuyas manifestaciones bioquímicas, celulares y fisiológicas pueden afectar profundamente a algunos datos de la investigación. Ya sea que los animales estén ubicados en nuevas colonias de reproducción, o en una colonia en espera de ser asignados a un estudio, se debe considerar tiempo suficiente para este periodo de adaptación. Parece que es particularmente importante proporcionar periodos de adaptación para los animales que van a ser estudiados en condiciones de sujeción.

Resumen

Los factores ambientales que afectan al bienestar de los primates en hábitat cautivos comienzan con un compromiso de tipo institucional hacia el bienestar de los animales y el adiestramiento del personal que los atenderá y usará. Luego se elabora un plan de acción, en el que se indica quién será el responsable de los animales, a quiénes se les confiere autoridad para hacer determinadas tareas y qué tareas son esas, y una declaración de la meta global. El bienestar de los primates no humanos no se puede garantizar si no se toman estas medidas iniciales. Después es preciso determinar los tipos y configuraciones de la vivienda con base en un conocimiento de las especies y los fines del programa. Deben formularse estrategias para el desarrollo de grupos sociales, y para retirar progenie o animales individuales. El personal debe ser seleccionado y adiestrado específicamente para que conozca las especies con que ha de trabajar, y se le dará tiempo suficiente para aprender las diferencias individuales de modo que se familiarice tanto con los comportamientos de los miembros de la colonia, que pueda reconocer fácilmente cualquier cambio sutil. Cada persona debe saber la medida de sus limitaciones y cuándo pedir ayuda a otros. El asesoramiento de especialistas externos se debe solicitar como cuestión de rutina. El bienestar y el rendimiento reproductivo de los primates no humanos solo pueden asegurarse con personas conocedoras que estén motivadas para atender las necesidades del animal. Muchos tipos diferentes de viviendas y estrategias de crianza pueden llenar estas necesidades, pero solo las personas idóneas pueden asegurar su éxito.

Conservacion de Primates no Humanos y su Importancia para la Salud Publica

Robert A. Whitney, Jr.

Director del Centro Nacional de Recursos para Investigaciones Institutos Nacionales de Salud
Washington D.C.
EUA

Los primates no humanos son indispensables en la investigación biomédica moderna, en la producción de biológicos y en la prueba de compuestos para determinar su toxicidad. Son especialmente valiosos por su parentesco evolutivo con el hombre, no solo por la similitud anatómica general y de comportamiento, sino por sus condiciones fisiológicas y bioquímicas análogas a las humanas. Debido a esta estrecha relación, los primates ofrecen una comprensión muy particular de situaciones paralelas en el hombre. Los primates no humanos han sido no solo la clave en el desarrollo de la vacuna contra la poliomielitis, sino que también han contribuido a ampliar aun más el conocimiento y la comprensión de otras enfermedades como la malaria, la fiebre amarilla, el sarampión, la tuberculosis, los trastornos mentales y la oncogénesis vírica, entre otras.

Actualmente, se lleva a cabo un gran número de importantes proyectos de investigación biomédica en los laboratorios del mundo. Además, la aplicación de los resultados de las investigaciones depende en alto grado de la utilización de primates no humanos. Sin realizar una prueba preliminar en estos animales, los riesgos de la aplicación del conocimiento teórico en los humanos pudieran ser demasiado graves.

La diseminación rápida y repentina del síndrome de inmunodeficiencia adquirida (SIDA) ha puesto aun más en evidencia la importancia de los primates no humanos en la investigación biomédica. Cabe recalcar una vez más que es su estrecha semejanza con los humanos lo que los convierte en modelos tan valiosos. Como se sabe, el virus de la inmunodeficiencia humana (VIH) es, casi con certeza, un virus mutante de los simios. Varios retrovirus símicos que producen enfermedades parecidas al SIDA en ciertas especies de monos, pero que en otras no producen ninguna enfermedad, desempeñan una función importante en el desarrollo de medicamentos y vacunas contra el VIH. Los chimpances, propensos a ser infectados por el virus humano causante del SIDA, aunque no presentan síntomas clínicos de la enfermedad, son esenciales para probar las vacunas humanas potenciales contra el SIDA, como lo han sido y lo continuarán siendo en la investigación de la hepatitis B.

El brote del SIDA es desvastador en sí mismo, pero sirve también de potente advertencia de que no sabemos qué problemas de salud pública puedan surgir en el futuro, cuando los modelos de primates no humanos pudieran ser indispensables para obtener una respuesta.

En la época de Pasteur y Koch, cuando la investigación biomédica experimentó su gran desarrollo, la mayor parte de los estudios se realizaban en animales y los adelantos importantes que se lograron fueron, a menudo, el resultado directo de esos experimentos. Naturalmente, en aquel momento el interés se centraba en los animales más asequibles. Sin embargo, aun entonces se llevaban a cabo estudios con primates no humanos. El Dr. Lebert of Breaslau fue probablemente el primer investigador que utilizó monos como sujetos de sus estudios sobre transmisión de enfermedades. En 1869, comenzó la descripción patológica de la tuberculosis en los monos que morían en el Zoológico de Breaslau.

Durante la primera mitad del siglo XX, los primates se convirtieron lentamente en modelos animales de laboratorio. El mono rhesus (*Macaca mulatta*) fue el que se utilizó más comúnmente, aparentemente porque era hasta cierto punto fácil de conseguir, de tamaño conveniente para la mayoría de los estudios y relativamente resistente. Las características particulares de ciertas especies las hacen especialmente valiosas en algunos estudios específicos, en especial los monos del Nuevo Mundo. Los hallazgos relacionados con la oncogénesis vírica en los primates requieren varios monos de este tipo. Las investigaciones sobre las cepas resistentes a los medicamentos contra la malaria dependen de la utilización del mono musmuqui (*Aotus* spp.), susceptible a las cepas humanas de la malaria. El mono pichico boca blanca (*Saguinus mystax*) es susceptible a la hepatitis vírica humana A, y es importante en el desarrollo tanto de procedimientos diagnósticos como de una vacuna.

Hasta el decenio de los años setenta casi todos los primates que se requerían en la investigación biomédica se obtenían de su habitat natural. Un estudio llevado a cabo en 1978, cuando se experimentó una gran escasez de primates, demostró que la comunidad biomédica de los Estados Unidos necesitaba anualmente 34.000 primates no humanos aproximadamente, de 35 diferentes especies. Más de dos terceras partes eran primates que procedían del Viejo Mundo y la mitad de estos, macacos. El mono verde y el mandril africanos formaban el resto de las especies del Viejo Mundo

En cuanto a los monos del Nuevo Mundo, se determinó que la necesidad mayor era de monos frailecitos (*Saimiri sciureus*), seguido del pichico boca blanca (*Saguinus* spp.), del musmuqui (*Aotus*) y del mono tití común (*Callithrix jacchus*). Se necesitaban muchas otras especies en cantidades más reducidas, para ciertos proyectos específicos de investigación. (En esta presentación me refiero a todos los *Saguinus* como "pichicos", limitando la palabra "tití" al *Callithrix* spp., de acuerdo con el uso taxonómico actual).

Los monos musmuqui y los pichicos son especialmente valiosos en la investigación de algunas enfermedades específicas importantes. Los monos frailecito son modelos animales versátiles que se pueden utilizar en una amplia variedad de estudios. De manera que no es sorprendente que doce años después de esos estudios tales especies de primates del Nuevo Mundo sigan siendo las más utilizadas en la investigación biomédica. Los monos musmuqui, valiosos en la investigación de la malaria, también son usados en los estudios sobre el glaucoma y la visión, y en la investigación sobre el virus respiratorio sincitial en la bronconeumonía y la bronquitis. Los pichicos son imprescindibles en la investigación de la hepatitis A, y también importantes en el estudio de otras enfermedades, especialmente en los campos de la inmunología, la virología y la oncología. El *Saguinus oedipus* es valioso en la investigación del cáncer del colon y la colitis. Los animales que se utilizan en la actualidad proceden todos de centros de reproducción, ya que esta especie amenazada no se pueden obtener del bosque natural.

Los monos frailecitos son valiosos en campos tan variados como la investigación del comportamiento, la malaria, las enfermedades cardíacas y la reproducción, el desarrollo y la prueba de medicamentos, e investigaciones sobre vuelos espaciales. Asimismo, se requieren cantidades más pequeñas de otras especies, ya que constituyen modelos convenientes para ciertos estudios específicos.

Cuando se concluyó el estudio de 1978, la destrucción del hábitat de los primates, la captura excesiva para la venta como mascotas y la caza de subsistencia amenazaban a las poblaciones naturales de primates, como hoy en día. Aunque la pérdida de primates por la demanda de la investigación fue y es muy pequeña en comparación de la enorme destrucción de poblaciones, las medidas adoptadas por los gobiernos de los países han afectado también inevitablemente a la investigación biomédica. En 1973, la India redujo en 50% la exportación de monos rhesus y en 1978 prohibió totalmente la exportación de primates. El mono más utilizado en los laboratorios, el rhesus, se hizo inasequible. Brasil, Colombia, Perú y otros países de América Latina, preocupados por la conservación, han asimismo vedado la exportación de primates.

Entre los diversos programas formulados para proteger la salud pública internacional mediante el aseguramiento del abastecimiento continuo de primates no humanos para uso biomédico, me concentraré en aquellas iniciativas basadas en el Plan Nacional de Primates de los Estados Unidos. El estudio de 1978 mencionado anteriormente fue el punto de partida de ese plan, desarrollado por el Comité Interagencial para el Manejo de Primates, bajo la presidencia del Subsecretario de Salud, quien dirige el Servicio de Salud Pública de los Estados Unidos. Los miembros del Comité representan a los diferentes organismos del Gobierno de los Estados Unidos encargados de la salud pública y de otras actividades relacionadas. Este Comité continúa hoy en día con una misión más amplia y un nuevo nombre: Comité Interagencial para Investigaciones con Animales. Yo actúo de presidente de ese Comité. Debido a las necesidades especiales de primates no humanos para la investigación y las pruebas biomédicas, una de nuestras principales preocupaciones es que se disponga de un abastecimiento permanente y suficiente.

Los tres pasos principales del Plan Nacional de Primates son las estipulaciones fundamentales para la conservación de los primates no humanos necesarios en la salud pública, bajo cualquier patrocinio. Esos tres pasos son:

- ampliar la producción de primates en los países usuarios;
- desarrollar programas internacionales para la producción y conservación de varias especies de primates, y
- hacer uso más efectivo de los primates que se dispone.

En los Estados Unidos el mayor desarrollo en la producción local ha sido, naturalmente, la del mono rhesus, que casi no se podría obtener de otra manera. La crianza local tiene otras ventajas además de la de evitar la escasez. Los animales deberían tener mejor calidad y condiciones más uniformes. Los centros de reproducción de monos libres de patógenos específicos, hoy en día raros, aumentarán, acrecentando aun más el valor de estos animales en la investigación de las enfermedades virales. El Centro Nacional del Instituto Nacional de Salud de Recursos para Investigaciones de los Institutos Nacionales de Salud (EUA) está desarrollando actualmente colonias de reproducción de monos rhesus libres de patógenos específicos, con el fin de proveer mejores modelos para su utilización en la investigación del SIDA.

Los comerciantes no han desarrollado centros de reproducción de primates del Nuevo Mundo en los Estados Unidos, debido a que la investigación requiere cantidades más pequeñas de estos animales. La reproducción local se lleva a cabo entre la comunidad de investigadores, tal como los centros de reproducción del mono musmuqui en el Instituto Battelle, el Centro Regional de Investigaciones sobre Primates (New England), el Instituto Militar de Investigación Walter Reed, y el centro de recursos de investigación y reproducción del mono frailecito boliviano en la Universidad de Alabama, en Mobile, Alabama.

En este último centro, la investigación se concentra en la ampliación del conocimiento en cuanto a la biología del mono frailecito. Esta investigación tiene dos metas: primero, mantener mejor en cautiverio los monos frailecito, de manera de no perderlos debido a la destrucción de su hábitat, y segundo, hacer que la investigación biomédica con estos monos sea más eficiente y productiva, de manera que se requiera una cantidad menor de ellos.

En cuanto al segundo punto sobre el desarrollo de la producción internacional y los programas de conservación, el programa de cooperación técnica de la OPS para la conservación, reproducción y utilización de primates no humanos ha sido evidentemente la iniciativa más importante basada en el plan. Este será el tema de la próxima presentación en el día de hoy. Señalaré solamente que el establecimiento y desarrollo del Proyecto Peruano de Primatología, incluyendo el Centro de Reproducción de Iquitos, fue un paso de tremenda transcendencia tanto para la conservación de las especies de primates en referencia como para la salud pública mundial, especialmente en lo que respecta a las enfermedades infecciosas.

La Organización Panamericana de la Salud era, sin duda, el punto focal adecuado y efectivo para iniciar esos esfuerzos de cooperación internacional tan esenciales para la salud pública mundial. La función de la OPS continua siendo crucial para satisfacer las necesidades de primates no humanos en el campo de la salud pública, a medida que continúa la destrucción de su hábitat.

El tercer elemento, la acción para asegurar la utilización más efectiva de los primates que se dispone, tiene muchas facetas. Entre ellas: la de no utilizar primates cuando exista otro tipo de animal que pueda producir resultados comparables; no utilizar aquellas especies de primates que existen en poca cantidad, cuando se disponga de otras especies más abundantes que resulten apropiadas; utilizar el mínimo número de primates que produzcan resultados científicos aceptables, y practicar la utilización múltiple y el reciclaje humanitario y científicamente compatible.

Aunque los dos primeros pasos son importantes, los centros locales de crianza y la utilización efectiva de los animales ya disponibles no proveerán primates en suficiente cantidad para satisfacer las necesidades mundiales de salud pública en la investigación biomédica, producción de biológicos y pruebas de toxicidad. La conservación y el manejo de las especies de primates en su hábitat natural son y continuarán siendo de gran importancia.

A medida que surjan nuevas tecnologías útiles, es probable que aumente la demanda de especies que no se utilizan mucho en la actualidad. No sabemos y no tenemos manera de saber qué modelos de primates se necesitarán en el futuro, especialmente en el campo de la inmunología. El descubrimiento reciente de un retrovirus en los monos verdes africanos, que les es innocuo pero que resulta mortal para otras especies de monos del Nuevo Mundo, puede crear situaciones paralelas en especies que actualmente no se utilizan en la investigación.

En vista de la destrucción acelerada del hábitat de los primates no humanos desde que se estableció el Plan Nacional de Primates de los EUA, el enfoque de tres fases establecido en ese plan cobra más importancia hoy en día. En beneficio de toda la humanidad, debemos insistir en la construcción de centros locales de reproducción, en los esfuerzos internacionales para que los países que han constituido una fuente de abastecimiento de primates continúen proporcionándolos, y en la utilización eficaz y precavida de esos recursos irreemplazables.

COOPERACION TECNICA DE LA OPS/OMS EN CONSERVACION Y UTILIZACION DE LOS PRIMATES NO HUMANOS EN LAS AMERICAS

Primo V. Arámbulo III

Alfonso Ruíz

ORGANIZACIÓN PANAMERICANA DE LA SALUD
Oficina Regional de la Organización Mundial de la Salud
Washington, D.C. 20037

Los cambios ocurridos en la distribución y movilización de los poblados humanos han acarreado la destrucción de los hábitats de la población de primates no humanos, causando la grave reducción de su población. Como resultado, existe un creciente interés en el desarrollo de programas dirigidos a conservar a los primates salvajes como un recurso natural renovable, que se pueda manejar y utilizar para beneficio mutuo del hombre y los animales. Los primates no humanos desempeñan una importante función en la salud pública. Ellos son un factor clave en el desarrollo de vacunas, y han contribuido de manera extensa a ampliar el conocimiento y la comprensión de numerosas enfermedades. La Organización Panamericana de la Salud, Oficina Regional de la Organización Mundial de la Salud para las Américas, reconoce, por medio de su programa general de cooperación técnica en salud pública, la función esencial de los primates no humanos. La OPS colabora activamente con los Países Miembros en el desarrollo y fortalecimiento de actividades para la conservación, reproducción y utilización racional de los primates no humanos mediante su Programa de Salud Pública Veterinaria.

La rápida expansión de los asentamientos humanos hacia las áreas naturales vírgenes y el desarrollo agroindustrial de varios países han sido las causas principales de la destrucción de los bosques, que han ocasionado la reducción de la población de primates salvajes y la disminución del número de especies nativas.

La concientización cada vez más creciente en cuanto a la valiosa función que tienen los primates no humanos en la salud pública ha tenido eco en los gobiernos de varios países, que han procedido a restringir el comercio de primates, algunas veces prohibiendo, su exportación. No obstante, esta acción no ha resuelto aún el problema básico de la destrucción de los hábitat. Afortunadamente, algunos países han mostrado interés en llevar a cabo programas tendientes a preservar los primates salvajes como recursos naturales renovables para fines biomédicos.

Como consecuencia de esta inquietud, los Países Miembros de la OMS, durante las XXVIII y

XXIX Asambleas Mundiales de la Salud, manifestaron en forma colectiva su preocupación por el riesgo de desaparición de los primates no humanos a causa de la destrucción de sus hábitat. En estas Asambleas se aprobaron las resoluciones WHA28.3 y WHA29.67 respectivamente, que piden al Director General de la OMS: "que facilite el intercambio tanto de recursos como de tecnología entre todos los países interesados y, con la ayuda de otras organizaciones internacionales interesadas según proceda, ponga el asesoramiento de expertos a la disposición de los países, y a solicitud de estos, para conservar, criar, suministrar y utilizar primates no humanos".

Mediante la ejecución de este mandato se espera que los países procedan a elaborar programas nacionales con los siguientes objetivos:

- Asignar atención especial a las actividades de conservación, con el fin de que los primates se desarrollen de manera permanente en su hábitat natural;
- Manejar la población natural de primates como recurso renovable;
- Instar a los Gobiernos Miembros, con poblaciones naturales de primates,el control de la cosecha, bien sea la captura periódica en la selva o en las colonias locales de reproducción,
- Lograr un efecto positivo en el desarrollo comunitario dentro de las áreas designadas para conservación.

Tomando en cuenta estos objetivos se han establecido las siguientes estrategias para desarrollar programas de conservación, reproducción y utilización racionales de primates no humanos en las Américas.

- Conservación y reproducción en sus hábitat naturales.
- Reproducción en cautiverio.
- Cooperación entre países para desarrollar y fortalecer programas nacionales de primatología.
- Desarrollo de tecnología adecuada para promover la utilización de los recursos autóctonos.

El problema principal que se enfrenta para desarrollar los programas nacionales y regionales sobre primates no humanos es el dilema que presenta la implicación de dos aspectos antagónicos:

1. La conservación de primates.
2. La utilización racional de primates para fines biomédicos.

En junio de 1975, la Organización Panamericana de la Salud y el Gobierno del Perú convocaron la "Primera Conferencia Interamericana sobre conservación y utilización de primates no humanos neotropicales en la investigación biomédica". Esta Conferencia recomendó la creación de un grupo internacional, que incluyera a la OPS y la Unión Internacional para la Conservación de la Naturaleza y los Recursos Naturales, con el objeto de que se estudiaran las prioridades para desarrollar colonias de reproducción de primates en cautiverio, se establecieran zonas naturales de reserva para proteger las especies autóctonas y se elaborara una caracterización taxonómica de las especies de primates. Asimismo, recomendó a todos los países ratificar la Convención sobre el Comercio Internacional de Especies en Peligro de Extinción (CITES).

A mediados de los años setenta la OPS ha apoyado la elaboración de programas nacionales de primatología con los gobiernos del Brasil, Bolivia, colombia y Perú. Los avances con Colombia se vieron interrumpidos en el desastre de noviembre 1985 que arrasó la Estación de Cría de Armero, con la muerte de casi toda la colonia de *Aotus* y del investigador responsable. En Brasil el esfuerzo principal fue orientado hacia el entrenamiento y la capacitación para la reproducción en cautiverio en el "Centro Nacional de Primates" del Ministerio de Salud en Belem. El Programa Primatológico del Perú iniciado en 1976 ha evolucionado hasta convertirse en una buena organización que combina el manejo en áreas naturales y la crianza en cautiverio, desde la sede en Iquitos donde se ubica el Centro de Reproducción y Conservación de Primates.

En 1976, el Gobierno del Perú promulgó una política nacional para la investigación y el manejo de la fauna silvestre, la cual considera el Proyecto de Primatología una de sus líneas de acción. La Organización Panamericana de la Salud (OPS) suministró la cooperación técnica necesaria para desarrollar y ejecutar el Proyecto Peruano de Primatología, llevando a cabo estudios de censo y dinámica de población para proporcionar las bases científicas para la cosecha controlada y la subsistencia de las especies.

Al desarrollarse el proyecto, la OPS formuló objetivos más amplios, como son los siguientes:

- Formular recomendaciones para la protección de áreas silvestres adecuadas que garanticen la protección y conservación de las especies de primates no humanos.

- Obtener y suministrar la información básica apropiada para la planificación y ejecución de módulos de manejo para las poblaciones naturales y la crianza.

- Desarrollar métodos para la reproducción de primates no humanos tanto en cautiverio como en su hábitat natural, y promover su conservación en la selva.

- Determinar el estado actual, la ecología, distribución y dinámica de las poblaciones de los primates no humanos, especialmente aquellos que se necesitan en la investigación biomédica.

- Obtener información y elaborar métodos para estabilizar y mejorar el hábitat de especies seleccionadas de la flora y fauna silvestre.

- Colaborar con las autoridades nacionales en el desarrollo de planes de conservación y reproducción de primates no humanos, y ayudar en la movilización de los recursos financieros externos necesarios para su ejecución.

- Promover la utilización de los primates en las actividades y recursos programáticos como vehículo para fomentar el desarrollo comunitario integral en las áreas programadas.

La operación del proyecto requiere un marco de referencia de actividades bien coordinadas que comprendan tanto los intereses de los países que utilizan primates como los de los países que llevan a cabo programas de conservación y reproducción.

La figura N° 1 muestra los canales principales de la cooperación técnica global planificada por la OPS para los proyectos de conservación y reproducción de primates no humanos, la cual suministra experiencia técnica y de política a los países con programas de conservación y reproducción, así como también a los países que requieren primates para fines biomédicos; facilita la transferencia de primates de origen reconocido y maneja los ingresos por concepto de derechos de transferencia, los cuales se destinan a apoyar las actividades de conservación y reproducción.

Figura N° 1 DIAGRAMA ESQUEMATICO DE LA COOPERACION TECNICA DE LA OPS/OMS PARA LA CONSERVACION Y REPRODUCCION DE PRIMATES NO HUMANOS

1. Adoptar políticas nacionales para el desarrollo sostenido.
2. Cooperar en la preparación de planes y estrategias de acción para el programa nacional de primatología
3. Informar sobre los programas de investigación y la necesidad de primates
4. Movilizar los recursos necesarios para desarrollar programas de conservación y reproducción en los países de origen.
5. Informar a los gobiernos en cuanto a la disponibilidad de primates en exceso de la población natural o los que nacen en cautiverio.
6. Facilitar la transferencia de primates
7. Suministrar el pago por concepto de derecho de transferencia y conservación, e informar sobre los resultados de la investigación que se realiza.
8. Administrar los ingresos por concepto de derechos de transferencia para apoyar las actividades de reproducción y conservación.

La ejecución de proyectos de conservación y reproducción en los países, requiere una organización estructural concreta, formada especialmente de un comité nacional y un comité técnico. Estos dos comités están constituidos por representantes de todos los organismos involucrados en la coordinación, ejecución y evaluación de las actividades programadas. La ejecución de las actividades se realiza por medio de los centros para la conservación y reproducción de primates no humanos, que tienen tres programas: reproducción y salud, laboratorio y áreas naturales.

En lo que respecta al Proyecto Peruano de Primatólogia (PPP), el Ministerio de Agricultura, el Ministerio de Salud, la Universidad Nacional Mayor de San Marcos, el Instituto de Investigación de la Amazonía Peruana y la Organización Panamericana de la Salud conforman el Consejo consultivo que integra el Comité Nacional y comparte la responsabilidad del proyecto (figura N° 2 y N° 3).

Figura N° 2 PROYECTO PERUANO DE PRIMATOLOGIA ORGANIGRAMA

Figura N° 3 PROYECTO PERUANO DE PRIMATOLOGIA COMPOSICION DEL CONSEJO CONSULTIVO

PRESIDENTE:	- MINISTRO DE AGRICULTURA
MIEMBROS:	- MINISTERIO DE SALUD
	- ASAMBLEA DE LA REGION DEL AMAZON
	- UNIVERSIDAD DE SAN MARCOS (IVITA)
	- INSTITUTO DE INVESTIGACIONES DE LA AMAZONIA PERUANA
	- OPS/OMS
SECRETARIO EX OFICIO:	- DIRECTOR EJECUTIVO, CRCP

El Comité Técnico del PPP consiste de un Directorio presidido por la División de Agricultura de la Región del Amazonas, y constituida por los representantes de la Dirección de Salud de la Región del Amazonas, el Instituto de Investigación de la Amazonía Peruana, el Instituto Veterinario de Investigaciones Tropicales y de Altura (IVITA), la Organización Panamericana de la Salud/Organización Mundial de la Salud. (Figura N° 4)

Figura N° 4 PROYECTO PERUANO DE PRIMATOLOGIA COMPOSICION DEL DIRECTORIO

PRESIDENTE: DIVISION DE AGRICULTURA DE LA REGION DEL AMAZONAS

MIEMBROS:

- DIRECCION DE SALUD DE LA REGION DEL AMAZONAS
- INSTITUTO DE INVESTIGACIONES DE LA AMAZONIA PERUANA
- INSTITUTO VETERINARIO DE INVESTIGACIONES TROPICALES Y DE ALTURA (IVITA)
- OPS/OMS

El Proyecto Peruano de Primatología ha logrado resultados considerables, especialmente en lo que se refiere a la utilización racional de los primates no humanos, habiendo disminuido marcadamente el número de animales exportados, como se muestra en el cuadro N° 1.

Cuadro N° 1 PRIMATES NO HUMANOS ATRAPADOS EN EL PERÚ DURANTE 1964-1989

	EXPORTACIONES 1964-1974 (11 YEARS)	TRANSFERENCIAS A IQUITOS PERÚ 1975-1989 (14 YEARS)	%
No. animales	339,080	8,164	
X/Year	30,825	583	Decremento 97.6

La Organización Panamericana de la Salud y el PPP salvan de la exportación a 22.661 primates aproximadamente, cada año, en comparación con las cifras anteriores correspondientes al período 1964-1974.

De acuerdo con la ejecución del proyecto, se transfirió a la Organización Panamericana de la Salud un total de 8.164 primates, para su distribución entre los institutos de investigación biomédica. Las principales especies transferidas fueron el *Saimiri sp.* (34,2%), *Saguinus mystax* (33,3%) y el *Aotus sp* (20,4%), cuadro N° 2.

Cuadro N°2 PRIMATES NO HUMANOS TRANSFERIDOS A LA ORGANIZACION PANAMERICANA DE LA SALUD, 1976-1989 CENTRO DE REPRODUCCION Y CONSERVACION DE PRIMATES NO HUMANOS (CRCP), PROYECTO PERUANO DE PRIMATOLOGIA (PPP), IQUITOS, PERÚ, JULIO 1990

Especies	Total	%
Saguinus Mystax	2,715	33.3
Saguinus Fuscicolis	661	8.1
Saguinus Labiatus	155	1.9
Saimiri	2,793	34.2
Cebuella Pygmaea	150	1.8
Aotus	1,667	20.4
Cebus Apella	23	.3
Total	8,164	100

El Centro para la reproducción y conservación de primates no humanos en Iquitos ha alcanzado un progreso importante, teniendo actualmente una población reproductora que asciende a 751 primates, distribuida en ocho diferentes especies: *A. nacymae*, *A. vociferans*, *C. pygmea*, *S. fuscicollis*, *S. labiatus*, *S. mystax*, *S. boliviensis* y *S. sciureus* (Cuadro N° 3).

Cuadro N° 3 POBLACION DE PRIMATES EN EL CENTRO DE REPRODUCCION Y CONSERVACION DE PRIMATES NO HUMANOS, PROYECTO PERUANO DE PRIMATOLOGIA (PPP), IQUITOS, PERÚ, 1990

ESPECIES	ADULTOS		JUVENILES		HIJOS		TOTAL (%)
	M	H	M	H	M	H	
AOTUS							
A. NANCYMAI	82	82	33	38	22	18	275 (36.6)
A. VOCIFERANS	52	60	8	4	4	4	132 (17.6)
C. PIGMAEA	2	3	1	-	-	-	6 (0.8)
SAGUINUS							
S. FUSCICOLLIS	11	8	1	3	-	-	23 (3.1)
S. LABIATUS	7	5	3	3	-	-	18 (2.4)
S. MYSTAX	21	17	4	6	2	2	52 (6.9)
SAIMIRI							
S. BOLIVIENSIS	13	64	26	2	3	12	120 (16.0)
S. SCIUREUS	9	41	32	28	7	8	125 (16.6)
TOTAL	**197**	**280**	**108**	**84**	**38**	**44**	**751 (100.0)**

M = MACHO H = HEMBRA

El Proyecto Peruano de Primatología ha logrado otros beneficios que no son tan evidentes, particularmente en el desarrollo comunitario de su área de influencia. Esto ha contribuido a la atención primaria de salud de la población, la educación comunitaria y al enlace intersectorial para el ecodesarrollo, y al mejoramiento general del bienestar socioeconómico de la población humana que vive en las zonas selváticas.

EL MANEJO DE LA FAUNA EN BRASIL

Jordan Paulo Wallauer

Medico Veterinário
Divisão de Fauna e Flora Silvestre
Instituto Brasileiro do Meio Ambiente e dos Recursos Naturais
Renováveis - IBAMA
Brasil

Los objetivos que vamos a presentar son los siguientes:

a) La situación del manejo de la fauna en Brasil con especial atención en su evolución histórica.

b) El planeamiento para el manejo de la fauna en la actualidad brasileña.

c) Las tendencias del manejo de la fauna en Brasil.

d) La situación actual de la cría de primates neotropicales en Brasil.

Historia

Durante todo el tiempo de la colonización de Brasil los recursos naturales renovables, entre ellos los faunísticos fueron explotados como recursos inagotables; causa y consecuencia de la formación cultural que aún se percibe actualmente. Con efecto se comprueba la utilización de animales silvestres como alimento y mascota y eso persiste no solo en las regiones de difícil acceso, sino también en las grandes metrópolis donde el comercio de la fauna silvestre es activo y significativo, siendo hoy en día, uno de los principales responsables por las amenazas de extinción de a algunas especies, entre ellas la mayor parte de las especies de primates de la Selva Atlántica.

Las principales razones, del desaparecimiento de la fauna en la costa brasileña y principalmente en la región del nordeste, fueron las modificaciones ambientales provocadas durante la colonización por las plantaciones de caña de azúcar y café. Ese fenómeno es, aún en los días actuales, el responsable por la presión sobre los recursos forestales y faunísticos en la Región Centro-Oeste (soya, maíz y arroz) y en las márgenes y a lo largo de los ejes camineros de la Amazonía.

Mientras tanto se debe tener en cuenta que las fronteras de expansión agrícola, en los últimos 40 años, corresponden en su extensión a aquellas ocupadas en los 450 años anteriores, desde el descubrimiento del Brasil. Todo eso por culpa de las tecnologías empleadas para el deforestamiento y cultivo del suelo.

La caza, originalmente era tan solo un medio de obtención de alimento, lo que aún persiste en el medio rural. Mas adelante, se transformó también en deporte, reglamentado desde los años 30 a 60 por normas elaboradas sin ninguna fundamentación científica, lo que explica el hecho de que entre las especies consideras cinegéticas, figuraban psitacídeos, primates y otras interesantes para la práctica del deporte.

La casi desaparición de algunas especies y del aumento de las presiones que se ejercían sobre las demás creó una conciencia de conservación y por lo tanto fue el determinante para promulgación de la Ley 5.197 de enero de 1967, conocida como la Ley de Protección de la Fauna. Esa ley establece que todos los animales silvestres son patrimonio del Estado y que es prohibida su utilización, persecución, caza o captura; prohíbe la caza profesional, autoriza el comercio de productos y sub-productos de animales silvestres provenientes de criaderos y permite la caza deportiva, bajo determinadas condiciones.

Tan solo después de la creación de la Ley 5.197, el Estado se fue preparando para aplicarla. Así mismo de manera lenta y gradual, tanto en términos de estructura administrativa, como en el desarrollo y captación de conocimientos técnicos y científicos.

Al final de los años 1970 surgieron las primeras reglamentaciones que definieron los criaderos previstos en la Ley y se establecieron las condiciones para su funcionamiento. Inicialmente los reglamentos fueron restrictivos, pero poco a poco perfeccionados, al paso que incorporaban nuevos conocimientos surgidos del medio científico, hasta el punto de que posibilitarían la existencia de unos pocos criaderos de fauna silvestre con finalidad comercial. Esos reglamentos también posibilitaron la cría de animales silvestres con fines de investigación científica y que también tuvo como resultado la legalización de colecciones privadas de fauna silvestre con producción científica reducida.

En relación a la caza deportiva, esta solo se ejerce en el Estado del Rió grande do Sul, en el extremo sur del país, con base en levantamiento periódicos de las poblaciones de especies cinegéticas, y la fijación de cuotas de piezas por cazador y establecimiento de temporadas anuales. La caza para subsistencia es más dañina que la deportiva, principalmente en las áreas de mayor ocupación humana, ya que la practican todo el año, incluso en las épocas de reproducción, que son las de veda.

Hasta mediados de los años 80, teníamos un modelo de manejo de fauna mucho mas consecuente con los hábitos culturales, las demandas de las actividades ambientales y de la necesidad de soluciones urgentes para problemas graves, que fruto del conocimiento de la dinámica de las especies faunísticas y sus reales necesidades en lo que respecta a su protección o su potencial para explotación, o aún más acerca del nivel técnico del Estado para la comprensión de su propio papel como administrador los recursos naturales del país.

El planeamiento para el manejo de la fauna

Tan solo en los últimos tres o cuatro años empezó en las organizaciones públicas responsables por la política y administración de los recursos faunísticos, la necesidad de planificar las acciones a medio y largo plazo, así como para dar respuesta a los problemas graves buscando atender las demandas de los distintos segmentos de la sociedad, a veces contradictorias. Desde luego se buscaran directrices, se instituirán reglas técnicamente más coherentes y se discutirán con más desenvoltura la real problemática del manejo de la fauna y su relación con los aspectos socio-económicos del país. Como consecuencia de esto, es posible en el Brasil, la recría de tortugas la de Amazônia (podocnemis expansa). Aún es posible recriar caimanes del Pantanal (Caiman crocodilus yacare) manejando huevos de esas especies, sacados de la naturaleza.

Esas actividades fueron reglamentadas en 1989 y 1990 en base a los datos científicos desarrollados en el IBAMA, particularmente en el caso de la tortuga de la Amazonia, y por una Universidad y un empresario rural, en el caso de los caimanes.

Aún se estudia una manera de mejorar técnicamente los lepitópteros en base al incremento de la población nativa, por intermedio del cultivo de plantas que sirven le de alimento, en pequeñas áreas bajo control, y recria de sus formas larvales, como lo determina la ley.

Ese cambio de comportamiento de la administración pública es motivada por una serie de factores, entre de ellos:

1. La evolución del conocimiento técnico-científico y mayor experiencia del personal encargado de la administración de los recursos faunísticos.

2. La existencia de directrices globales desarrolladas por las organizaciones internacionales como la IUCN la FAO y ONU que generaron documentaciones tales como: La Estrategia Mundial para la Conservación y el informe de la Comisión Mundial acerca del Medio Ambiente y Desarrollo (Nuestro Futuro Común).

3. La participación de entidades tales como la "World Wildlife Foundation", la "Conservation International" y otras más, en programas de conservación de la fauna en Brasil.

4. Mayor participación de la iniciativa privada, como interesada en las posibilidades económicas que se puede obtener con el uso sostenido de la fauna silvestre.

5. Participación de los técnicos en programas de entrenamientos y eventos en el exterior, como fue el caso de los "Seminarios Latino Americanos para gerentes de vida silvestre", promoción del U. S. Fish and Wildlife Service o de los cursos del CATIE, en Costa Rica.

Sin embargo, algunas dificultades siguen siendo enfrentadas por la administración pública, que intenta encontrar las soluciones para los problemas relacionados con la ecología, flora y fauna e intenta al mismo tiempo, disminuir los gastos públicos. Lo que determinó significativos cortes presupuestarios, principalmente con gastos de personal. Esas dificultades siguen siendo atenuadas buscando la participación cada vez mas, la iniciativa privada y las Organizaciones no Gubernamentales.

Específicamente en lo que se refiere a la utilización de primates para las investigaciones de la medicina, se puede decir que se busca la participación de los laboratorios y institutos de investigación privados, para el desarrollo de programas y proyectos de conservación y uso sostenido de esos animales. Como ejemplo podemos comentar sobre dos hospitales brasileños especializados en el tratamiento de la lepra, los cuales obtuvieron su registro del IBAMA, (órgano responsable para la política y administración del medio ambiente y recursos naturales renovables) y desarrollaron la cría en cautiverio de una especie de armadillo (Dasipus novencintus) para la obtención de la mitsudina, necesaria para el diagnóstico del Mal de Hansen.

Tendencias

Las tendencias de la polltica de manejo de la fauna a corto y medio plazo son de planificar acciones, que corresponden a las directrices orientadas el uso sostenible de esos recursos naturales y a la manutención de la biodiversidad, teniendo como estrategias:

1. La protección de las especies amenazadas de extinción, tanto por medio de la protección de sus habitas, como por el establecimiento de bancos genéticos ex situ, lo que se hace con los primates del genero Leonthopithecus. Este banco genético participa en la cría de esas especies fuera del país, como en los Estados Unidos y en New Jersey, en Gran Bretaña entre otras.

2. El desarrollo de modelos de cría en cautiverio con base en el manejo de huevos e hijos - al ejemplo de lo que se permite para el caiman del Pantanal y la tortuga de Amazonia.

3. El incentivo a la cría con finalidad comercial para las especies no amenazadas de extinción - capibara (*hidrochaerus hydrocheris*), coypus (*myocastor coypus*), el pichico común (*callithris jacchus*), entre otros.

4. El desarrollo de estudios acerca de la viabilidad de caza deportiva para las especies con potencial para ello - dentro de esas se citan algunas tinamifomes, anatídeos y columbídeos.

5. La autorización para la cría con finalidad científica para los animales considerados amenazados de extinción, de acuerdo con las leyes brasileñas o incluso en la CITES. Esas crías, sin fines lucrativos, podrán obtener animales para investigaciones médicas, pero deben intentar alcanzar, principalmente el conocimiento de la biología y el desarrollo de las técnicas de cría en cautiverio.

6. La utilización de los excedentes de los zoológicos - lo que, en algunos casos representa centenas de especies como lo que ocurre con el mono (Cebus apella).

7. La utilización de la educación y de la extensión para cambiar hábitos culturales, creando una conciencia positiva y adecuadas al potencia de uso de los recursos faunísticos.

A lo largo del tiempo y dependiendo del suceso obtenido con las estrategias anteriormente mencionadas, se puede observar con antelación que la investigación necesitara modelos de uso sostenido de poblaciones de animales silvestres y que podrá servir para las modificaciones de la Ley 5.197, así como posibilitar la legalización del manejo con explotación in situ, actualmente prohibido. Tal manejo, con certeza es tan solo posible desde una concientización adecuada de los beneficiarios en todos los niveles, y del desarrollo de sistemas de control y fiscalización.

La cría de primates neotropicales en Brasil

La situación actual de los criaderos de primates presentan un cuadro en el cual incluye diez registro como criaderos científicos y uno con finalidad comercial. En esos criaderos oficializados para la finalidad de investigación, son críadas las especies:

Callithrix aurita
Callithrix jacchus
Callithrix penicillata
Callithrix seoffroyi
Callithris flaviceps
Callithris humeralifer
Callithrix kuhlii
Callithrix argentata
Leontopithecus rosalia rosalia
Leontopithecus r. chrysomellas
Leontopithecus r. chrysopysus
Cebuella pysmea
Sasuinus bicolor
Sasuinus midas
Saimiri sciureus
Callicebus moloch
Chiropotes satanas
Aotus sp. (2 especies)

Cebus apella (3 especies)
Cebus albifrons
Brachyteles arachnoides

Comercialmente son creadas las especies *Callithris jacchus* e *callithris seoffroyi,* con una explotación de 60 C. *jacchus* e 14 C. *seoffroyi* en el año de 1990 y una expectativa de explotación de 120 C. *jacchus e* 40C. *seoffroyi* para 1991.

Igualmente, aunque no registrado como criadero, la Fundacâo Parque Zoológico de Sâo Paulo realiza un trabajo muy bueno de cría de tres subespecies de *lontopithecus rosalia*, en integración con el Centro de Primatología de Rió de Janeiro, Zoológico de los Estados Unidos de América y con el Jersey Wildlife Preservation Trust, en un programa para la preservación de esos primates.

Hay que destacar los experimentos del Centro de Primatología de Rió de janeiro con híbridos de especies de *Callithris*, en búsqueda de un camada mas resistente, fértil y adaptado al cautiverio, para la utilización en la investigación científica en general.

No obstante el suceso en la reproducción de casi todas las especies, principalmente con las del genero *Callithrix,* los criaderos con finalidad científica, vinculados a la administración pública, se encuentran, actualmente en dificultades para mantener sus planteles y ampliar sus investigaciones como es el caso del Centro de Primatología de Universidad de Brasilia, actualmente casi desactivado. Para mantenerlos produciendo trabajos que, benefician a toda la humanidad, por su esfuerzo en la manutención de la biodiversidad y por sus investigaciones de interés médico, será necesaria también la cooperación de la iniciativa privada y de las entidades no gubernamentales.

Manejo de los Primates No-humanos como Recurso Renovable: la Economia de la Conservacion

Theodore Panayotou

Harvard Institute for International Development,
Kennedy School of Government,
Harvard University,
Cambridge, Massachusetts 02138, USA.

Primo V. Arámbulo III

ORGANIZACIÓN PANAMERICANA DE LA SALUD
Oficina Regional de la Organización Mundial de la Salud
Washington, D.C. 20037

La presencia ubicua de primates no humanos en los zoológicos ha ayudado a aumentar su reconocimiento como una fuente de diversión y de educación para el público en general. Sin embargo, sigue subestimándose su importancia en otras áreas tales como las de la salud pública, la investigación biomédica, el equilibrio ecológico y el sustento de las poblaciones locales. Tradicionalmente, los primates no humanos en los neotrópicos eran capturados como fuente de alimentos y de ingresos. La carne de mono se vendía en los pueblos y ciudades de todo el Amazonas y los animales vivos eran comercializados activamente como mascotas y animales experimentales para investigación biomédica.

En ciertas áreas, tales como el Amazonas del Perú, la captura excesiva y la destrucción del habitat natural han conducido a un agudo agotamiento de la población de primates silvestres. En respuesta, el gobierno del Perú ha prohibido, desde 1973, la captura, retención y comercialización de primates no humanos en tanto se otorgan permisos especiales para la captura de primates con fines científicos. En un esfuerzo por conservar y utilizar racionalmente la población de primates, los países con una población de primates importante, en colaboración con la Organización Panamericana de la Salud, han formulado políticas y proyectos para la gestión de los primates no humanos como recurso renovable. Un ejemplo pertinente es el Proyecto Peruano de Primatología que comenzó en 1972. Este proyecto considera a los primates no humanos como un recurso renovable con el cual, mediante la debida gestión, se podrían satisfacer las necesidades de la comunidad de investigación biomédica en tanto se asegura la supervivencia a largo plazo de estas especies en su habitat natural.

Este documento tiene por finalidad analizar los conceptos y principios de la economía de recursos renovables potencialmente pertinentes para la conservación y utilización racional de los primates no humanos. No se hace un intento por aplicar estos principios a casos específicos ni por obtener estimaciones científicas. Analizamos sucesivamente los conceptos del máximo rendimiento sostenible, el máximo rendimiento económico, los costos de oportunidad del habitat y la valoración de los enfoques para los primates.

Máximo Rendimiento Sostenible

Aun cuando cada especie tiene su propia dinámica de población, la mayoría de las poblaciones biológicas, incluidos los primates no humanos, tienden a seguir una curva lógica de crecimiento. Es decir, tienen un crecimiento bajo a niveles de población bajos y altos y un crecimiento alto a algunos niveles intermedios de población. Esto se debe a que el crecimiento es el resultado neto de la mortalidad y la reproducción que, a su vez, dependen de la dimensión de la población de primates en relación con la disponibilidad de alimentos. Con bajos niveles de población, el crecimiento es reducido ya que la reproducción es baja (debido al bajo número de animales de edad reproductiva) y la mortalidad es baja (debido a la alta disponibilidad de alimentos). Con altos niveles de población, el crecimiento también es bajo debido a que la reproducción y la mortalidad son elevadas (por razón de la baja disponibilidad de alimentos por animal). Libre de interferencias, una población de primates alcanzaría un equilibrio natural en el que el crecimiento neto de la población es de cero (véase la figura N° 1). Naturalmente, debido a fluctuaciones naturales en la disponibilidad de alimentos (por razón de una fructificación irregular) y en el área habitable (por razón de las inundaciones), la población de primates puede variar de un año a otro, incluso en ausencia de perturbación por el hombre, debido a una mortalidad más elevada (más baja) y/o a migración en años malos (buenos). Dichas fluctuaciones pueden ser considerables y deberían tomarse en cuenta en la gestión de los primates como recurso renovable.

Aun cuando podemos postular a priori que el crecimiento anual se maximiza a un cierto nivel de población intermedio en el que el número de animales reproductivos y la disponibilidad de alimentos son relativamente elevados pero no excesivos, el nivel exacto varía según las especies y el ecosistema y sólo puede estimarse empíricamente. En realidad, es más probable que éste esté representado por una gama de niveles de población en vez de por un sólo nivel debido a las fluctuaciones naturales en la disponibilidad de alimentos y en el habitat. Para sostener cualquier nivel dado de población, sólo deberán capturarse los animales producto del crecimiento o rendimiento anual, ni más ni menos. Si se capturasen más, la población descendería; si se capturasen menos la población aumentaría. Con un equilibrio máximo o una población máxima, el crecimiento y, por tanto, el rendimiento sostenible es cero. Para generar crecimiento, ha de capturarse parte de la población de la misma forma que ha de abrirse la selva clímax para que ocurra un crecimiento neto. El rendimiento máximo sostenible (MSY) se obtiene a un nivel de población que da lugar a un crecimiento máximo. En vista de la explotación excesiva ocurrida en el pasado, algunas poblaciones de primates pueden hallarse a niveles por debajo del nivel que corresponde al MSY.

Ciertas especies tales como el *Cacajao calvus rubicundus* y el *Lagothrix flavicauda* están en peligro, lo que significa que su población se ha reducido por debajo del umbral o nivel crítico mínimo (Xo); en otras palabras, sus números se han reducido a niveles que no aseguran la supervivencia. Otras especies, tales como el *Cebus albifrons cuscinus* y el *Aluata palliata* son vulnerables, lo que significa que sus números se están aproximando a niveles mínimos de supervivencia. Estos niveles de supervivencia críticos mínimos son específicos para especies individuales y deberían evaluarse conservativamente debido a la variación natural y a la posible irreversibilidad, una vez que se cruzan dichos niveles críticos mínimos. La gestión de los primates como recurso renovable requiere el estricto establecimiento y observación de estos niveles críticos mínimos puesto que la supervivencia es uno de los objetivos. Naturalmente, puesto que el uso racional requiere mantener la población de primates al nivel que genere el máximo rendimiento sostenible, existe poco riesgo de que la población se reduzca por debajo del nivel crítico mínimo de supervivencia con una gestión racional. Sin embargo, para las especies que tienen una población de supervivencia mínima muy grande, y para las que las tasas de reproducción son bajas y la disponibilidad de alimentos está sujeta a amplias fluctuaciones, las gestión de rendimiento máximo sostenible no es una garantía para la supervivencia.

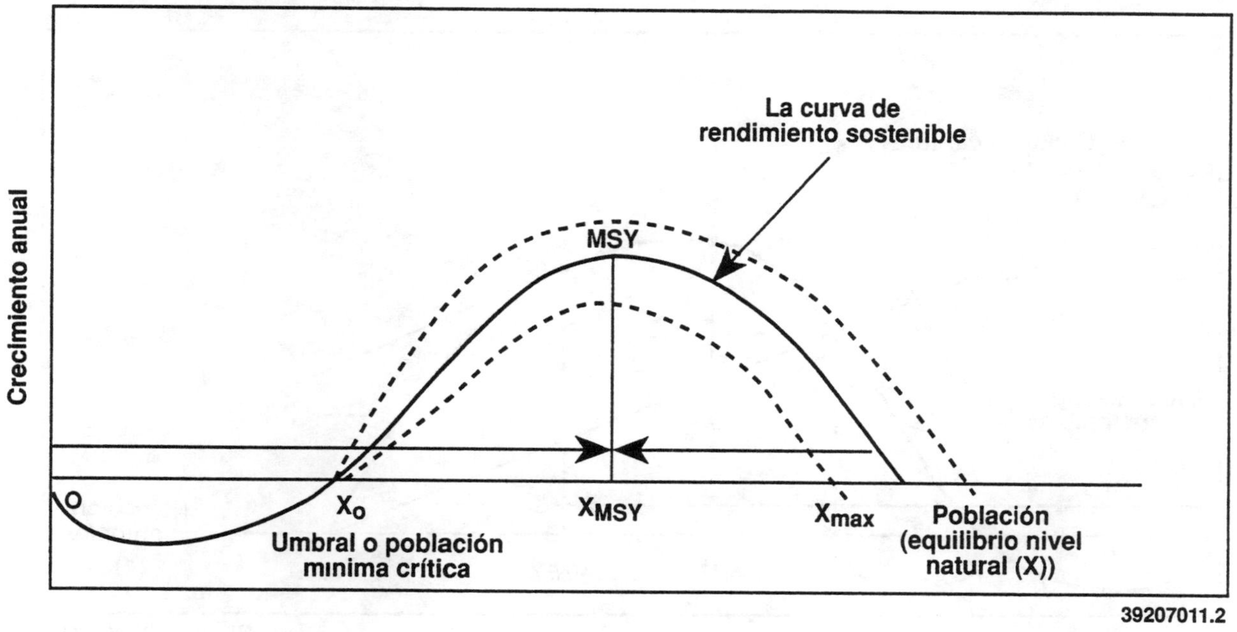

Figura N° 1: Gestión de las poblaciones de primates para un máximo rendimiento sostenible

Rendimiento Económico Máximo

Un objetivo de gestión biológicamente más conservador y económicamente más beneficioso es el rendimiento económico máximo (MEY). Este objetivo reconoce que, debido a que la captura y comercialización no están carentes de costo, no es rentable capturar el número de animales que representan el rendimiento máximo sostenible, de la misma forma que no es rentable aspirar a un rendimiento mínimo (en vez de óptimo) en cualquier actividad productiva. El ingreso o beneficio neto máximo se obtiene a un rendimiento sostenible más bajo (crecimiento), y por ende a un nivel de población más elevado, que el máximo rendimiento sostenible, tal como se indica en la figura N° 2.

La curva de ingresos sostenibles de la figura N° 2 es simplemente la curva de rendimiento sostenible de la figura N° 1, multiplicada por el precio de un primate, por ejemplo, vendido para fines de investigación biomédica o exposición en zoológicos. La curva del costo de captura y comercialización es el producto del número de primates capturados multiplicado por el costo de la captura, transporte y comercialización. Tal como se ve en la figura N° 2, la gestión de los primates para un rendimiento económico máximo (sostenible) mantiene en su habitat natural a una población de primates notablemente mayor que la gestión para un rendimiento (físico) máximo sostenible. Sin embargo, si se produce una explotación incontrolada en condiciones de acceso libre (OA), la población de primates descendería a XOA y el rendimiento económico se disiparía como resultado de la actividad de captura excesiva. Esta es la situación que prevaleció en el Amazonas del Perú con

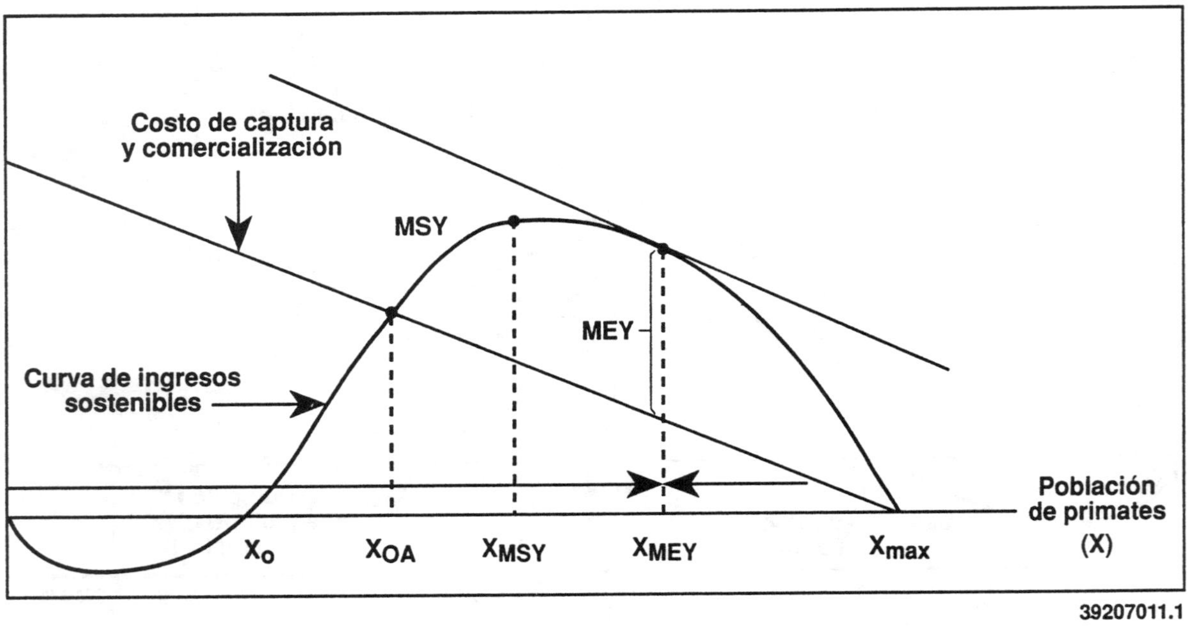

Figura N° 2: Gestión de las poblaciones de primates para un rendimiento económico máximo.

anterioridad a 1973 cuando la libre captura y comercialización de los primates fue prohibida y se otorgaron permisos para la captura controlada con fines de investigación biomédica.

Es fácil ver cómo la captura incontrolada con acceso libre pudiera conducir a un agudo agotamiento y a la posible extinción de los primates. Tal como se muestra en la figura N° 3, una combinación de reducción de la población de primates a un nivel bajo de XOA y condiciones naturales desfavorables (extensas inundaciones, fructificación errática o epidemias) pudieran reducir la población de primates por debajo de su nivel mínimo de supervivencia.

Así pues, la utilización racional de los primates como recurso natural renovable no sólo les protege de un agotamiento agudo y de la amenaza de extinción sino que, además, genera ingresos sustanciales para su protección y gestión, control sanitario, mejora de la infraestructura y desarrollo de la comunidad.

Costos de Oportunidad del Habitat

Si los recursos que entran en juego fuesen las pesquerías marinas, el análisis precedente habría estado completo salvo en lo que respecta al hecho de que optamos por no tratar los aspectos dinámicos (en cualquier caso, se prevé que el rendimiento dinámico máximo caerá en un punto comprendido entre el MSY y el MEY). Sin embargo, los primates difieren en cierto número de

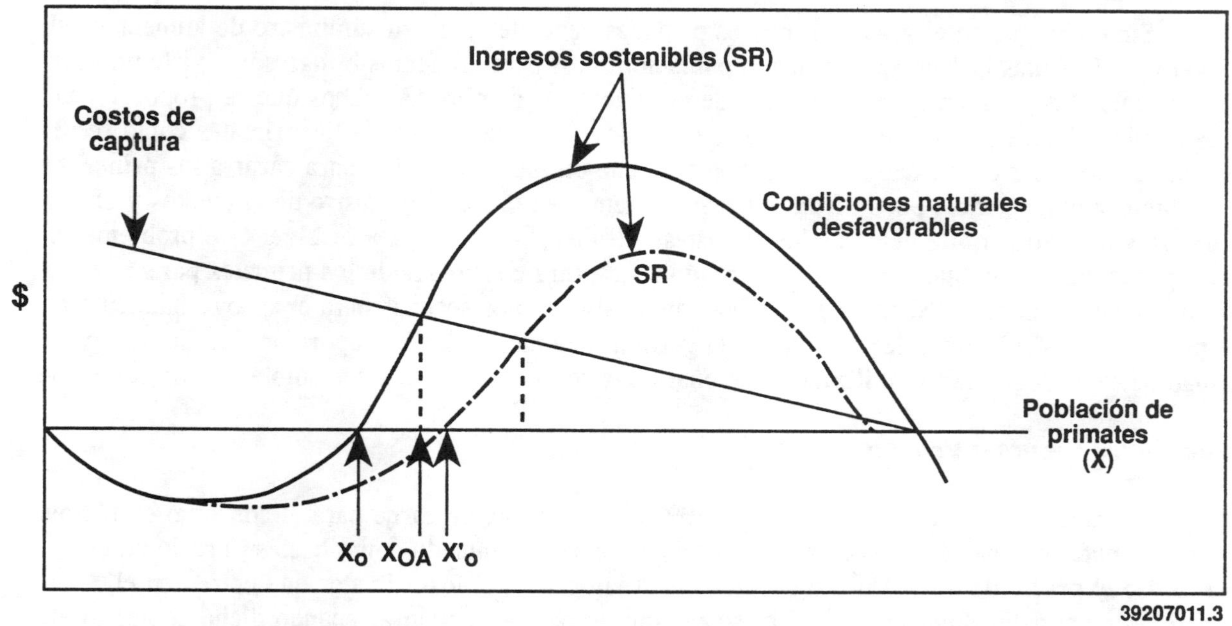

Figura N° 3: Agotamiento agudo y amenaza de extinción debido a una combinación de explotación incontrolada y condiciones naturales desfavorables.

formas de las pesquerías cuyo habitat marino no tiene a menudo un costo de oportunidad (salvo en las zonas costeras) y cuyo valor se limita de ordinario a los ingresos procedentes de la venta de las capturas (excepto en el caso de los arrecifes coralíferos donde entran en juego valores de biodiversidad).

El mantenimiento de una población más amplia de primates, como ocurre con el MEY, la gestión frente a la explotación incontrolada, requiere un habitat mayor, que es un área más extensa de selva bastante libre de perturbación, como fuente de alimentos y espacio vital. Si existe un uso de valor más elevado para el recurso forestal (por ejemplo, la madera) y para la tierra (por ejemplo, la agricultura), el mantenimiento del área bajo selva como habitat para los primates tiene un claro costo de oportunidad. Este es igual al valor más elevado perdido por la explotación maderera y la agricultura menos el valor de la selva en su estado natural (excluyendo el valor de los primates). En el caso del Proyecto Peruano de Primatología, tenemos la fortuna de contar aproximadamente en la misma zona con una valoración comparativa de una hectárea de terreno sometida a 3 usos de alternativa: ganadería, plantaciones de árboles y recolección de productos no madereros (frutas). En 1988 el valor actualizado de una hectárea de terreno en la zona del Amazonas del Perú cerca de Iquitos era de $9.920 para los productos forestales no madereros, de $3.184 para la madera y pulpa de madera y de $2.960 para la explotación ganadera. Por tanto, el mejor uso del terreno es el mantenimiento de la selva en su estado natural y su explotación para productos forestales no madereros. En este caso, no existen costos de oportunidad del terreno o la selva al utilizar racionalmente las poblaciones de primates en la misma zona como un recurso renovable para obtener un rendimiento económico máximo.

Sin embargo, en el grado en que los primates dependen para su suministro de alimentos del mismo tipo de frutas que las que utiliza la población local para alimentos e ingresos, existe un costo de oportunidad de mantener una población de primates más amplia. A menos que se proporcionen otras posibles fuentes de subsistencia para la población local, la gestión de los primates como recurso renovable puede ser difícil debido a que la gente tiene un incentivo doble para cazar a los primates : (a) reducir la competencia por la fruta y (b) suplementar su propio suministro de alimentos y obtener ingresos adicionales con la venta de los primates. El mejor enfoque para resolver este problema es hacer participar a la población local en la gestión y captura controlada de los primates para investigación biomédica. Si se da a la población local derechos seguros para el aprovechamiento de los primates (con sujeción a las normas de la gestión racional) y se les proporciona ayuda técnica, probablemente asegurarán la utilización racional de la reserva, de forma sostenible, mejor que nadie.

Valor de los Recursos Primatológicos

Cuando los primates se capturan y utilizan como fuente de carne para el consumo en el hogar o para la venta, su valoración podría realizarse simplemente tomando como base su precio en el mercado o el precio de una carne en sustitución. Análogamente, los primates que entran en el comercio de animales domesticables o mascotas fueron fáciles de valorar cuando dicho comercio era legal. Con la prohibición de las capturas, tenencia y comercialización de los primates por el Perú y otros países en la región, es más difícil obtener precios de mercado para los primates vendidos como mascotas. Sin embargo, los gobiernos y la OPS aspiran explícitamente a eliminar dicho comercio ilegal. La participación de la población local en la gestión y captura legal de los primates es un medio eficaz de hacerlo, siempre y cuando los ingresos por el comercio legal excedan a los del comercio ilegal menos una prima de riesgo, y estos beneficios se distribuyan ampliamente a la población local.

El valor de los primates comercializados legalmente para investigación biomédica y observación en zoológicos debería ser fácil de obtener de los gobiernos o grupos comerciales afectados. Sin embargo, se tropieza con dos dificultades: (a) el mercado legal para los primates puede ser demasiado reducido para que prevalezca la competencia entre los compradores y (b) las fuentes de determinadas especies apropiadas para investigación biomédica son escasas y, por tanto, hay una competencia inadecuada entre los vendedores. El Perú y Brasil probablemente tengan un monopolio de algunas especies. En términos de cría y manejo, los primates neotrópicos (concentrados en el Amazonas) son más fáciles de manejar en condiciones de laboratorio debido a su tamaño relativamente menor en comparación con sus homólogos del viejo mundo.

En los parques nacionales, santuarios de la fauna y reservas naturales, los primates constituyen una parte de la vida silvestre que atrae a los turistas. Cuanto mayor sea la población de primates, tanto más frecuentes serán los animales vistos y, por tanto, tanto mayor será la probabilidad de que los turistas visiten el lugar. La estimación del valor estético de los primates no es una tarea fácil debido a que las personas visitan los parques o reservas para ver a muchos animales y plantas, ríos, cascadas, etc. Necesitamos conocer la elasticidad de las visitas por los turistas (y sus gastos) con respecto a un primate adicional (u observación de primates) o evaluar de alguna forma el grado de preferencia de las personas (y su disposición a pagar) entre lugares con diferentes probabilidades de observación de primates, manteniendo todos los demás factores constantes. En ausencia de un tal lujo, un uso creativo del costo del viaje y enfoques de valoración contingentes, pueden ayudar a obtener algunas estimaciones aproximadas. Adviértase que, al contrario de los valores de utilización analizados con anterioridad, la observación y fotografía de los primates no son usos consuntivos, es decir, no disminuyen la población de primates. La adición de este valor a los usos consuntivos analizados antes tendería, marginalmente, a apoyar mayores poblaciones de primates.

Aparte de los valores de utilización arriba indicados, los primates tienen varios otros valores que son menos cuantificables. Primero, tenemos el valor ecológico de los primates: estos animales desempeñan un papel en la floración, fructificación, dispersión y propagación de semillas de la flora silvícola que contribuyen a las asociaciones entre las plantas y el equilibrio ecológico. Aun cuando no es posible atribuir un valor monetario extra a esta contribución, su importancia cualitativa indica que una población de primates algo mayor es socialmente mejor que la que justificarían los valores de uso solamente.

Los usuarios de un recurso, bien se dedique éste a un uso consuntivo o no consuntivo, tienden a dar un valor adicional al recurso expresado mediante opciones de utilización del recurso en el futuro como beneficio potencial distinto de su valor de utilización actualizado en el presente. Este valor de opción es una expresión de la preferencia y disposición a pagar un precio por preservar la opción de la disponibilidad y uso futuros del recurso. El valor de opción no se limita a los usuarios actuales; los no usuarios también pueden estar dispuestos a pagar algo por preservar la opción de uso futuro, si optan por ejercer esa opción. Así pues, las instituciones de investigación biomédica y los científicos, los zoológicos y el público observador y los visitantes actuales y futuros de los parques y reservas estarían dispuestos a pagar alguna cantidad positiva para asegurar que los primates seguirán estando disponibles para experimentos biomédicos, exposición en zoológicos y observación en parques y reservas. El método de valoración contingente se utiliza, de ordinario, para estimar los valores de opción. Este es un método de encuestas de acuerdo con el cual se pregunta a las personas si estarían dispuestas a pagar un precio por la conservación de los primates o si estarían dispuestas a aceptar una compensación por la pérdida de los primates. En ausencia de pagos reales, el problema de gratuitismo se resuelve mediante un diseño apropiado de la encuesta.

Segundo, los primates no humanos tienen una importancia estratégica ya que constituyen un fondo genético sumamente valioso que debería preservarse indefinidamente. Es una clase de valor de opción para la sociedad en su conjunto. Surge debido a la incertidumbre acerca de las necesidades y preferencias futuras y a la irreversibilidad de la extinción. En el caso de los primates no humanos, este valor probablemente sea muy elevado debido a su relación filogenética con el hombre. Hay importantes similitudes fisiológicas y de comportamiento que hacen que los primates sean modelos sumamente valiosos para la investigación biomédica. Algunos tienen condiciones de enfermedades que ocurren naturalmente y que son similares a las del hombre, lo que les hace útiles para estudiar las enfermedades humanas como sujetos experimentales en la prueba y control de la calidad de las vacunas y medicamentos. Este valor de opción social o, más bien, global es imposible de estimar y difícil de captar salvo mediante contribuciones internacionales. La ayuda proporcionada por la OPS y el NIH a los países con poblaciones de primates importantes, entre otros, son tales contribuciones.

Finalmente, los primates, como otros recursos renovables y ambientes naturales, tienen un valor intrínseco de por sí que, aunque difícil de definir y mucho más difícil de medir, se manifiesta en los numerosos fondos y organizaciones ambientales existentes para proteger las especies en peligro de extinción. La gente valora la existencia de un primate, incluso si no tienen proyectado utilizarlo directamente en la investigación, la observación o la fotografía (valor de existencia). La gente también desea asegurar la existencia de los primates, y de otras poblaciones de la fauna silvestre, como legado de generaciones futuras (valor de legado). Las personas también tienen una cierta simpatía por los animales, incluyendo los primates no humanos (por los que quizás tengan una simpatía especial), y están dispuestas a pagar algo por preservar su habitat y asegurar su supervivencia y bienestar. Cabe citar aquí las actividades de la OMS en respuesta a la preocupación internacional por reducir el uso de los animales en los laboratorios de investigación, hallar alternativas y establecer normas y disposiciones para el tratamiento humanitario de los primates no humanos. El valor de existencia puede estimarse utilizando el enfoque de valoración contingente tal como se explicó arriba.

Conclusión

La mayor amenaza para los primates no humanos es la destrucción de su habitat por la agricultura migratoria. La utilización racional de las poblaciones de primates no humanos como recursos renovables lograría el objetivo doble de la conservación y el uso racional que, en realidad, son dos lados de la misma moneda.

Por dos razones, la gestión para un rendimiento económico máximo (sostenible) es preferible a la gestión para un rendimiento máximo sostenible (físico): (a) mantiene una población más amplia que constituye un seguro contra las fluctuaciones naturales y límites críticos mínimos elevados de supervivencia; y (b) aumenta la cantidad máxima de ingresos para la protección de los recursos y desarrollo de la comunidad que proporcionarían una alternativa de subsistencia frente a la captura furtiva de primates por la población local. El medio más eficaz en función del costo para aplicar una política de gestión de recursos renovables consiste en hacer participar y habilitar a la población local en la gestión de los recursos.

Los costos de oportunidad de mantener una población más amplia de primates en los neotrópicos tienden a ser bajos si bien hay valores adicionales importantes, tales como los valores de uso para los turistas, los valores ecológicos, los valores de opción y de existencia. El nivel óptimo de población de primates y la tasa de aprovechamiento son una cuestión empírica que depende de la dinámica de la población de la especie en particular de que se trate, sus valores de utilización y no utilización y los costos de oportunidad del habitat y el aprovechamiento. La gestión de los primates no humanos como recurso renovable de un valor económico importante conservaría el recurso en beneficio del hombre y de los animales.

Desarrollo Integrado de Unidades de Conservacion y el Papel de los Programas Nacionales de Primatologia

Jose Vincente Rodriguez

Instituto Nacional de Los Recuros Naturales (INDERENA)
Renovables y del Ambiente
Bogota, Colombia

Colombia, que tiene una superficie de 1.141.784 km^2, está ubicada en la zona intertropical del noroeste de América del Sur. Tiene costas tanto sobre el mar Caribe como sobre el Océano Pacífico, y la atraviesa la línea ecuatorial. La mitad oriental del territorio, aproximadamente, consiste en llanuras bajas, que forman parte de las cuencas del Amazonas y el Orinoco. Esta zona se caracteriza principalmente por los afloramientos dispersos del Amazonas, así como mesetas y montañas, formaciones que están relacionadas geológica y biológicamente con la región montañosa de Guyana. Además, en la región sudoeste de la cuenca del Orinoco, junto a los Andes, se levanta otra cadena aislada: la Cordillera de la Macarena. En la mitad occidental del país hay tres grandes brazos de la Cordillera de los Andes: la Cordillera Oriental, que forma una cadena continua con los Andes venezolanos; la Cordillera Central, que es la prolongación septentrional de la Cordillera Real del Ecuador; y la Cordillera Occidental. En la costa del Caribe se encuentra la Sierra Nevada de Santa Marta, macizo montañoso enteramente aislado de la cordillera mencionada en el cual está ubicado el pico más alto del país (la Horqueta de Chundú, de 5.780 m de altura). Paralela a la parte norte de la costa del Pacífico se halla la Serranía del Baudo, y en la frontera con Panamá se encuentra la Serranía del Darién.

El clima es muy variado, ya que abarca una amplia gama de condiciones. Por ejemplo, en la península de Guajira, que constituye el extremo septentrional de América del Sur, predomina el clima caluroso desértico, con precipitaciones de 150 mm anuales en promedio, mientras que en los llanos del Chocó (llanura del Pacífico) las precipitaciones anuales alcanzan los 13.200 mm, en promedio. La temperatura disminuye con la altura, de manera que a unos 4.800 metros sobre el nivel del mar se encuentran nieves permanentes. La complejidad orográfica y climática, sumada a su situación geográfica y a una historia geológica bastante complicada, contribuye al desarrollo de un mosaico biótico muy complejo y quizás a la mayor biodiversidad del mundo, ya que en este país se encuentra alrededor del 10% de las especies de organismos vivientes.

Colombia tiene la mayor diversidad de primates de la región neotropical, con la excepción de Brasil. Las tres familias vivientes de Platyrrhini están representadas como sigue: Callithricidae (dos géneros con ocho subespecies), Callimiconidae (familia monotípica, con una especie) y Cebidae (nueve géneros con 19 especies). Se ha informado sobre la presencia de ejemplares de *Chiropotes satanas* en la ribera colombiana del río Orinoco, lo cual aún no se ha confirmado. Además, las siguientes especies, como mínimo, podrían encontrarse en Colombia: *Cacajao calvus*, *Aotus trivirgatus*, *Pithecia aequatorialis*, *Cebus nigrivittatus* y *Saguinus labiatus*. La mayoría de estas especies se encuentran en los bosques de los llanos, por debajo de los 1.000 metros sobre el nivel del mar, en el bosque pluvial, higrófilo (bosque húmedo), subhigrófilo (bosque semicaducifolio), higrotropófitico (bosque caducifolio) y en galería. La mayor diversidad se halla en el bosque del piedemonte amazónico, entre los ríos Caquetá y Putumayo. Varias especies, principalmente las de tamaño mediano y grande, llegan hasta el bosque andino, a una altura de 2.800 metros (*Cebus apella*) y, localmente, a los 3.000-3.200 metros (*Alouatta seniculus*, *Aotus lemurinus* y *Lagothrix lagotricha lugens*). Hay tres especies endémicas: *Saguinus oedipus* (*Sensu stricto*), *Saguinus leucopus* y *Aotus brumbacki*, así como otras dos especies de *Aotus* a las cuales todavía no se ha dado un nombre.

Al igual que en muchos otros países tropicales, el principal problema para la conservación de la fauna silvestre es la deforestación, ya que más de la mitad del bosque original ha sido destruido, particularmente en la mitad occidental del país, incluida la región andina que es la más poblada. La tala de bosques continúa debido a la gran presión para habilitar nuevas tierras para la agricultura y la ganadería. A ello se suma la presión de la caza principalmente como fuente de alimentos. Este problema se ve agravado por el hecho de que gran parte de las tierras desmontadas no son apropiadas para la agricultura sostenible o el desarrollo ganadero. Asimismo, grandes zonas con suelos de buena calidad se destinan a la ganadería intensiva de baja rentabilidad. Por otra parte, muchas de las tierras que se destinan a la agricultura intensiva tienen una clara vocación boscosa. Esta situación prevalente está poniendo en peligro a una parte creciente de la flora y la fauna silvestres de Colombia, lo cual requiere estrategias urgentes y eficaces de conservación y protección. En cuanto a la conservación de los primates no humanos del país, un programa de esta índole debería comprender, como mínimo, lo siguiente:

- La creación de parques nacionales y otras reservas y la consolidación de los existentes;

- Investigaciones básicas para complementar los conocimientos biológicos y ecológicos fundamentales de la especie de los primates, el estado de la población y un sistema adecuado de vigilancia para observar la población de primates y el estado de sus habitats.

- Sobre la base de dichas actividades, deberían llevarse a cabo proyectos orientados a la adquisición de conocimientos sobre los aspectos biológicos de la cría y otros aspectos decisivos que posibiliten la cría en cautiverio, tanto para conseguir animales con el fin de reponer las poblaciones naturales como para suministrar ejemplares para las investigaciones biomédicas.

- Este programa debería incluir también un componente de educación y divulgación, a fin de promover una mejor actitud de los seres humanos hacia los primates no humanos.

- Además, a fin de realizar estas actividades eficientemente, sería necesario establecer mecanismos de cooperación en términos de personal de proyectos de investigaciones, capacitación, equipo e instalaciones apropiados, asesoramiento científico y técnico, y asistencia financiera.

Actualmente, el Sistema Nacional de Parques de Colombia consiste en 41 zonas administradas por INDERENA, las cuales, junto con las reservas indígenas, abarcan más de 30 millones de hectáreas, de las cuales nueve millones pertenecen a distintas categorías del Sistema Nacional de Parques.

Si bien la superficie total de estas zonas es muy grande, lamentablemente estas reservas no son suficientes para proteger las 47 especies y subespecies de primates no humanos que se hallan en Colombia, ya que para ello habría que crear más reservas, y para contar con una representación apropiada de la gran variedad de ecosistemas, centros de endemicidad y elementos de la biota en peligro de extinción. Se presenta un análisis de los problemas y las soluciones relacionados con estos temas.

Asimismo, se formulan varias sugerencias y comentarios para la elaboración de la estrategia de conservación necesaria.

La Cooperacion Multisectorial e Interpaises Para el Desarrollo de Programas Nacionales de Primatologia

Manuel Uceda Castillo

Dirección General de Forestal y Fauna del
Ministerio de Agricultura del Perú
Lima, Perú

y

Presidente del Comité de Projecto Peruano de Primitologia "Manuel Moro Sommo"
Iquitos, Perú

Señores:

Es muy honroso tener oportunidad de dirigirme a ustedes, gracias a la gentil invitación de la Organización Panamericana de la Salud/Organización Mundial de la Salud, para exponerles nuestra propuesta sobre cooperación multisectorial e interpaíses para el desarrollo de Programas Nacionales de Primatología.

El Perú como Estado miembro de la Organización Mundial de la Salud tiene el compromiso de participar en esta Reunión, pues considera que su aporte habrá de constituir una alternativa para aliviar la salud humana, la preservación de nuestros recursos naturales y a través de ello, la posibilidad de alcanzar nuestro desarrollo.

Desde fines de 1972, debido a la preocupación mundial por la conservación y uso racional de los primates no humanos, el Gobierno Peruano, a través de sus Ministerios de Agricultura, Salud y Relaciones Exteriores, así como el Instituto Veterinario de Investigaciones Tropicales y de Altura (IVITA y con el apoyo decidido de la Organización panamericana de la Salud/OMS, dieron inicio a los primeros trabajos de investigación, evaluación y manejo de fauna silvestre en la Amazonía Peruana y bosques tropicales del departamento de Tumbes en la Costa Norte de nuestro país, recayendo en la Dirección General de Forestal y Fauna del Ministerio de Agricultura la responsabilidad de orientar y supervisar la ejecución del Proyecto.

En 1975, se consolida este interés mediante el funcionamiento del Proyecto Peruano de Primatología "Manuel Moro Sommo" en mérito a la suscripción de una Carta-Convenio con la Organización Panamericana de la Salud/OMS, la cual luego de tres enmiendas, ha extendido su vigencia hasta Diciembre del presente año. Como es evidente se prevé la renovación de la Carta-Convenio, luego de una evaluación de lo ejecutado, así como de las características y fines nobles - que cumple el Proyecto, en pro de la salud mundial.

El objetivo fundamental del Proyecto es la investigación, orientada a la conservación y manejo de los primates no humanos en el Perú, y para su cumplimiento se han desarrollados tres líneas de acción: a) Evaluación e investigación; b) Conservación y manejo; c) Educación, capacitación y extensión en el área de distribución de estas especies de mamíferos; con el propósito de promover la participación de las poblaciones locales, en la conservación de este recurso natural.

Del Proyecto podemos señalar algunos logros y beneficios que se han obtenido, entre ellos:

- Se ha llegado a identificar nueva especies de primates no humanos para el Perú, alcanzando actualmente un registro de 32 especies, así como importante información acerca de su distribución, estatus de conservación, ecología, etología y lineamientos para su manejo y aprovechamiento sostenido.

- Establecimiento oficial de un Zoocriadero denominado Centro de Reproducción y Conservación de Primates, en la ciudad de Iquitos, lo que ha posibilitado obtener mayor información y experiencia sobre manejo, crianza y reproducción en cautiverio de unas siete especies de primates de uso biomédico. Además cuenta con una biblioteca especializada que brinda servicio a la comunidad científica.

- Aprovechamiento de islas de la cuenca del Río Amazonas para la introducción de algunas especies de primates como el *Saguinus mystax* y *Saguinus labiatus* y su consiguiente manejo en semicautiverio habiéndose - obtenido notable éxito en el incremento de las poblaciones.

- Se ha logrado la comprensión y participación de las comunidades isleñas dentro de las acciones de manejo de primates en semicautiverio en las islas, a través de la protección de los primates; del desarrollo de la agricultura migratoria, y otros beneficios que otorga el Proyecto hacia la Comunidad en coordinación con otros sectores públicos, como Salud, para la atención de sanidad de pobladores locales; y Educación, mediante campañas de extensión y capacitación, así como en apoyo con infraestructura.

- Protección de especies de primates amenazadas de extinción o en situación vulnerable a causa de la caza furtiva y principalmente, la destrucción de su hábitat, consecuencia de la deforestación para fines agropecuarios.

  En este sentido, además de Parque Nacional del Manu, Reserva Nacional Pacaya-Samiria y otras Unidades de Conservación en la Región de la Selva, se ha propuesto la creación de las Reservas Comunales Tamshiyacu-Tahuayo (Dpto. Loreto) y del Tahuamanu (Dpto. Madre de Dios); así como el traslado de primates hacia áreas reservadas, a fin de garantizar su supervivencia.

- Formación de un equipo de profesionales y técnicos del más alto nivel en el campo de la Primatología, cuyos aportes son reconocidos por la Comunidad Científica Internacional.

- Contribución con los centros de investigaciones científicas a nivel mundial, muy particularmente en el área de los estudios biomédicos, mediante la transferencia de 8,405 especímenes de primates seleccionados de las especies: *Saguinus mystax* (2,745), *Saguinus fuscicollis* (661), *Saguinus labiatus* (255), *Saimiri sp.* (2,872) *Aotus sp* (1,699), *Cebuella pygmae* (150), *Cebus apella* (23). Esto es de gran relevancia, pues nuestros monos son excelentes modelos para producción de vacuna contra la hepatitis viral, la malaria, cáncer, el SIDA, entre otros males que afectan gravemente a la humanidad.

El Proyecto Peruano de Primatología, con casi 18 años de trabajo continuo, constituye un modelo para el manejo de la fauna silvestre, sus logros y beneficios son el resultado de la dedicación y esfuerzo coordinado de instituciones profesionales y técnicos nacionales e internacionales y de la Comunidad en su conjunto.

El Proyecto se desarrolla en base al documento "Estrategias y Plan de Acción Nacional 1985-1990 del Proyecto Peruano de Primatología".

A nivel regional, el encargado de coordinar las acciones y responsable de la ejecución del Proyecto y del cumplimiento de los acuerdos y políticas emanados del Consejo Consultivo, es el Directorio, con sede en la ciudad de Iquitos.

En cuanto a la ejecución del Proyecto, ésta se hace a través de sus Direcciones de Administración y de Investigación y Manejo.

No obstante esta organización, se procura establecer acciones coordinadas con los Sectores Educación, Pesquería, Interior, entre otros, así como con organizaciones no gubernamentales (ONG) y Comunales.

Ultimamente, se dieron algunas dispositivos legales; sin embargo, debemos - destacar el referente a la categorización de especies de fauna silvestre según su estado de conservación, mediante R.M. No. 01082-90-AG-DGFF del 14 SET.90; y el Código del Medio Ambiente y los Recursos Naturales, mediante Decreto Legislativo No. 611 del 07 SET. 90.

Paralelamente, a éste y otros esfuerzos, se han cumplido importantes coordinaciones con los países vecinos, con los que compartimos un patrimonio común y como logros tales acciones, se tiene los siguiente:

- Acuerdo para la Conservación de la Flora y Fauna de los Territorios Amazónicos de la República del Perú y de la República Federativa del Brasil suscrito el 07 NOV: 75 y aprobado por Decreto Ley No. 21670 del 26 OCT. 76.

- El Acuerdo para la Conservación de la Flora y Fauna de los Territorios Amazónicos de la República de Colombia y de la República Peruano, suscrito el 30 de Marzo de 1979.

- Creación del Comité Intergubernamental Técnico para la Protección y Manejo de la Flora y Fauna Amazónicas en la Primera Reunión Técnica sobre Parques Nacionales, Manejo y Conservación de la Biota Amazónica (Leticia, 28-30 Agosto 75.), resultado de una feliz iniciativa del Gobierno de Colombia durante la Reunión de Ministros de Agricultura en Cali, en 1975, habiendo asistido representantes de Brasil, Colombia, Perú, Venezuela y Bolivia.

El Gobierno del Perú tiene plena conciencia que la salvaguarda, aprovechamiento sostenido y defensa de esos recursos de flora y fauna silvestres, en particular los primates no humanos, que constituyen base sólida para el desarrollo armonioso de la Región, tienen que llevarse en acción conjunta entre los países de América, productores de primates neotropicales, con los que compartimos un patrimonio común y con los que nos unen los mismos intereses de desarrollo estable - para nuestros pueblos, así como de fines nobles en pro de la salud humana y animal. Es por ello que con gran complacencia vemos esta Reunión, en la que los funcionarios de los Gobiernos de la Región, discuten las acciones comunes que se deberán implementar para la protección y el manejo de los primates no humanos en particular, y de la fauna y flora neotropicales en general, recursos que bien sabemos han sido agredidos y que reclaman la mayor atención para concurrir a impulsar el bienestar general.

Con este propósito, a modo de contribución al debate que habrá de sostenerse, me permito proponer las siguientes recomendaciones que estimo pueden constituir adecuados canales de colaboración entre nuestros países.

1. Que el Comité Regional de Primatología para las Américas (CORP) sólo debería estar conformado por los países de América poseedores naturales de poblaciones de primates no humanos.

La Presidencia del Comité Regional debería recaer en uno de los representantes de los países poseedores de recurso, mientras que la Secretaría podría ser ejercida por la representación de la Organización Panamericana de la Salud.

2. Que para la operatividad de este Comité Regional, se recomienda la conformación de una Comisión que elabore y proponga el Proyecto de Reglamento de Organización y Funciones de este Organismo; a ser aprobado en Asamblea General.

3. Que los países miembros del CORP se comprometan a utilizar sólo primates provenientes de zoocriaderos o de áreas naturales bajo manejo, debidamente establecidos y/o controlados por el Estado.

4. Que se debe estimular y apoyar la ejecución de investigaciones biomédicas en los países productores de primates.

5. Que se mantenga la prohibición para la posesión, transporte y comercialización de primates como mascotas, así como sus restos y mantener un estricto control del tráfico de primates mediante el reforzamiento de los servicios fronterizos y la aplicación de la convención CITES.

6. Que, el CORP puede constituir un excelente canal de información permanente sobre los avances en las investigaciones y manejo, en los programa o proyectos nacionales de primates no humanos, así como para el intercambio de profesionales en la materia y el Centro de Reproducción y Conservación de Primates de Iquitos podría constituir la sede del Centro de Investigaciones Primatológicas de Latinoamérica.

7. Que, los países miembros del CORP realicen las gestiones necesarias con el propósito de lograr se establezcan laboratorios de investigaciones biomédicas en los países en los que se hace la extracción de primates.

8. Que, se recomiende a los países miembros del CORP para que se reserven áreas destinadas a la protección en su hábitat de especies de primates amenazados de extinción, y recomendar ante la OPS/OMS, para que ésta solicite a los centros de investigaciones y laboratorios que utilizan primates neotropicales, para que los productos biológicos finales obtenidos, se reviertan en beneficio de los países productores del recurso.

9. Recomendar la suscripción de convenios Binacionales entre países productores de primates con el propósito de reforzar la gestión del CORP y promover la conservación de los recursos de flora y fauna silvestres.

10. Que todo convenio que suscriba la Organización Panamericana de la Salud y que comprometa los recursos primatológicos de un país miembro, cuente con el oval y la visación de representante oficial del respectivo país.

11. Que se debe estudiar los mecanismos más adecuados para el financiamiento de las reuniones y operatividad del CORP a través de la OPS/OMS.

12. Se debe alentar la participación de los Sectores Público y Privado, así como de las agencias no gubernamentales para la instalación de zoocriaderos y manejo en semicautiverio de especies amenazadas y/o de alto valor científicos, cultural y/o comercial. Al respecto, convendría estudiar sus implicancias con relación a los programas estatales en funcionamiento.

13. Promover la participación de los pobladores locales en el desarrollo de los Proyectos y Programas de Primatología, a fin de que sean ejecutores y beneficiarios de los aportes que puedan generar tales acciones.

14. No se debe escatimar esfuerzos orientados a la ejecución de campañas de extensión y capacitación a todo nivel y por todos los medios de difusión, con el objetivo de relievar la importancia socioeconómica científica, cultural, ecológica y estratégica de los primates en particular y de los recursos de flora y fauna silvestres en general.

En suma, los representantes ante el CORP no debemos olvidar que los lineamientos de política que orientan las acciones de conservación de los primates se deben basar prioritariamente en asegurar la supervivencia e incremento de la población de las especies, en manejar racionalmente las poblaciones de primates y en beneficiar al trabajador del campo, a los profesionales nacionales, a los centro de investigación y de difusión cultural, y por ende, a todos los países miembros del CORP.

Señores representantes, esta I Reunión del CORP, a la cual dedicamos nuestros esfuerzos es de alta significación para la preservación de nuestros recursos y un medio para el mejoramiento de la calidad de vida de todos los habitantes de la Región, y porque no decirlo de la salud mundial.

Por ello, permitaseme, a nombre de mi Gobierno expresar mi reconocimiento a los señores representantes de los países miembros del Comité Regional de Primatología para las Américas CORP, a la Organización Panamericana de la Salud OPS/Organización Mundial de la Salud OMS y al Gobierno y Pueblo de los Estados Unidos de América, por su amable acogida.

GRACIAS

MOVILIZACION DE RECURSOS PARA LA CONSERVACION DE ECOSISTEMAS NEOTROPICALES: "CONVERSIONES DE DEUDA POR NATURALEZA"

Edward Liebow

Battelle Human Affairs Research Centers
Seattle, Washington
Estados Unidos

Introducción y resumen

Los enfoques posibles para la preservación de las poblaciones de primates neotropicales tienen tres conjuntos generales de objetivos: (1) la conservación de los hábitat; (2) la protección de las poblaciones de primates; y (3) la educación del público. Esta exposición presenta un enfoque específico a la conservación de los hábitats, la "Conversión de Deuda por Naturaleza" aunque debe reconocerse que ningún obstáculo específico impide la aplicación de las conversiones de deuda por naturaleza a los programas de protección de las poblaciones de primates o de educación del público. Esta idea fue planteada por primera vez en 1984, por el entonces vicepresidente del Worldlife Fund (Fondo por la Vida Silvestre Mundial) con sede en Estados Unidos. Hasta la fecha se han llevado a cabo 15 transacciones de este tipo y 9 de ellas en América Latina y el Caribe. Desde 1987, las conversiones efectuadas en las Américas han generado más de US$50 millones en fondos para fines de conservación; en Costa Rica (US$42,3 millones), en Ecuador (US$10 millones), Bolivia (US$250.000) y en la República Dominicana (US$600.000). Hasta la fecha, entre los principales grupos norteamericanos que han participado en estas conversiones se encuentran el World Wildlife Fund, Conservation International (Conservación Internacional) y Nature Conservancy (Conservación de la Naturaleza). El gobierno de Estados Unidos está considerando dar apoyo directo a este tipo de transacción, pero la autorización específica todavía está pendiente.

Observado superficialmente, este mecanismo innovador es fácil de comprender: una parte considerable de la deuda comercial de los países de América Latina y del Caribe fue contraída con instituciones crediticias de América del Norte, Japón y Europa, y parece poco probable que muchas de esas deudas puedan pagarse en su totalidad. Con las conversiones de deuda por naturaleza, es posible condonar a una nación deudora una parte pequeña de su deuda externa y establecer un fondo por intermedio de una transacción financiera escogida entre varias opciones que permite generar ingresos para apoyar la búsqueda de los objetivos de conservación. Sin embargo, la sencillez del esquema, contrasta con las complejas interacciones que han contribuido al violento ataque contra los bosques tropicales. Los países en desarrollo con grandes deudas externas han intentado sistemáticamente aumentar sus exportaciones para ganar dinero en efectivo y así cancelar sus préstamos. En su mayoría, estos esfuerzos se han realizado en detrimento de los bosques tropicales, cuyos árboles han sido talados para obtener madera o para dejar terrenos libres destinados a cultivos que generen dinero rápidamente. Los esquemas de deuda por naturaleza ofrecen incentivos para evitar esta práctica y promover las prácticas de desarrollo sostenible a la vez que se protege el valioso hábitat.

El esquema se basa en tres principios fundamentales para la acción:

- Para que un programa de deuda por naturaleza sea exitoso es necesario que en el país deudor haya instituciones reconocidas y fuertes dedicadas a la conservación; y

- Para tener éxito, los programas de deuda por naturaleza deben ser justos y además ser considerados como justos, es decir, todas las partes interesadas deben ser tomadas en consideración.

- Desde el punto de vista del país acreedor, los donantes filantrópicos e instituciones de crédito privadas son movilizadas, y su participación continua dependería de si se les da alguna seguridad de que los fondos que ellos donan están logrando su máximo efecto.

Con esta finalidad -- si se da cierta seguridad a los donantes de los países acreedores -- esta exposición recomienda algunas transferencias de tecnología que ayuden a sistematizar la consideración de los factores socioeconómicos junto con los criterios biofísicos utilizados para seleccionar las estrategias de conservación apropiadas.

Mi meta es describir brevemente los mecanismos de las conversiones de deuda por naturaleza, recordando que cada transacción individual tiene características propias y es compleja desde el punto de vista financiero. Luego abordaré brevemente el dilema que se crea en la intersección de dos aspectos relacionados con datos técnicos, valores del hábitat y tenencia de la tierra. Este dilema plantea la pregunta: ¿Cómo seleccionar estratégicamente las tierras adecuadas que deben destinarse a la conservación de los hábitats, dado que las tierras que tienen mayor valor desde el punto de vista de los hábitats pueden presentar problemas, en cuanto a propiedad o a patrones de uso de la tierra, difíciles de resolver, mientras que las tierras que está disponibles más fácilmente para fines de conservación quizás no tengan un valor significativo desde el punto de vista del hábitat.

Forma de efectuar una transacción de este tipo: los cinco pasos para una conversión de deuda por naturaleza

Las partes interesadas en la conversión suelen incluir a: (1) Las autoridades normativas en materia fiscal y financiera del país acreedor; (2) las instituciones de crédito del país acreedor; (3) la organización de conservación del país acreedor y las entidades que la respaldan financieramente; (4) autoridades financieras del país deudor (por ejemplo, banco central) y las autoridades que administran los recursos (por ejemplo, ministerios de agricultura y de la fauna silvestre); (5) las organizaciones de conservación del país deudor; y (6) las comunidades locales potencialmente afectadas por los programas de conservación. Cada parte tiene intereses diferentes (y a veces en conflicto) en la conversión, por lo tanto los arreglos para programas individuales de conversión deben ser hechos expresamente para tener en consideración los intereses específicos de todas las partes. Sin embargo, estos arreglos suelen efectuarse en cinco pasos:

1. Obtener la aprobación, en principio, del país deudor para el financiamiento de los mecanismos y las metas del programa de conservación. Junto con las partes del país acreedor que entran en un convenio, hay otras tres partes clave del país deudor que participan en las negociaciones: el gobierno, el banco central y una organización de conservación privada apropiada que recibirá los fondos y administrará el programa aprobado. La organización de conservación puede ser la parte interesada más importante de las tres, puesto que las organizaciones de conservación de los países desarrollados deben renunciar al control eficaz de los fondos para evitar el problema del "imperialismo", pero deben asegurar a los donantes y las autoridades fiscales el uso continuo y responsable de los fondos. Solo organizaciones privadas de conservación fuertes y reconocidas del país deudor pueden reunir este doble conjunto de condiciones.

Las negociaciones entre estas tres partes pueden abarcar una amplia variedad de aspectos, que incluyen entre otros, la tasa de cambio a aplicar para convertir la deuda en moneda local, las condiciones de pago o los bonos que pueden emitirse a cambio y cómo diseñar el programa de conservación al cual se aplican las ganancias.

2. **Obtención del instrumento de la deuda.** En este paso, la organización de conservación del país acreedor obtiene los fondos para adquirir el instrumento de la deuda en el mercado comercial. Hay un mercado secundario establecido para la deuda de los países en desarrollo que fijará un precio básico para la deuda del país. Los precios del mercado varían considerablemente, desde unos cuantos centavos de dólar hasta la casi paridad. Mientras menor sea el precio, mayor será el valor para efectuar una conversión de deuda por naturaleza, pero el precio del mercado también es un indicador de los riesgos en cuestión. El dinero necesario para la compra de deuda puede obtenerse de muchas maneras. Para las organizaciones de los países desarrollados, se trata de un problema clásico de movilización de fondos e inevitablemente trae a consideración si tal programa se apoya en nuevas fuentes de financiamiento o plantea un conflicto con otros esfuerzos de movilización de fondos. También debe recordarse la situación fiscal de las donaciones.

3. **Transferencia del título de la deuda.** Esta es una transacción compleja. Una decisión importante es saber quién será el comprador real. Puede resultar adecuado que la organización de conservación del país acreedor adquiera la deuda y luego la done a su socio en el país deudor. Puede ser posible donar los recursos necesarios para permitir la adquisición directa de la deuda por la organización de conservación del país deudor. Considerando un dictamen reciente del Departamento de Tesorería de los Estados Unidos, también es posible que la deuda sea donada directamente a la organización de conservación en el país deudor, que actúa como agente para la organización del país acreedor. Los impuestos y otras consideraciones financieras determinarán esta decisión.

4. **Conversión de la deuda de conformidad con el acuerdo logrado con el gobierno del país deudor.** Puede incluir la emisión de bonos en moneda local, la adopción de medidas para proteger ciertos aspectos difíciles (por ejemplo, compra de tierras o legislación protectora), y el pago de dinero en efectivo en moneda local para el rescate de la deuda. Los bonos son útiles ya que puede prorrogarse su fecha de vencimiento, generan ingresos para programas de despegue, y, al vencimiento, pueden convertirse en una donación para actividades futuras.

5. **Ejecución del programa de conservación acordado.** Esta es la meta final del programa, y tiene que ser vigilada muy cuidadosamente. Al mismo tiempo que se buscan los objetivos de conservación, puede estar en juego la reputación de la organización de conservación del país acreedor porque debe estar en capacidad de asegurar a los donantes y a las autoridades fiscales que los fondos recibidos se están empleando apropiadamente.

Valor del hábitat y tenencia de la tierra

Para que las naciones deudoras puedan obtener beneficios para el medio ambiente de estas transacciones, es fundamental la selección de áreas críticas de hábitats. Sin embargo, es igualmente importante entender los arreglos locales de tenencia de la tierra en su contexto cultural, de tal forma que la administración de tales áreas para fines de conservación de los hábitats evite una avalancha de consecuencias socioeconómicas adversas, lo que haría que los gobiernos deudores se mostraran renuentes a considerar tales conversiones.

La evaluación de los hábitats se basa, por supuesto, en características tales como la biodiversidad, la estabilidad, la intromisión de especies no autóctonas y la sensibilidad ante el entorpecimiento de las actividades humanas. La distribución de los valores del hábitat suele representarse en mapas identificando aquellas áreas que requieren atención especial de manejo de los recursos. Y aun allí donde se han emprendido esfuerzos de conservación, la mayoría de las áreas destinadas a conservación (por ejemplo, parques nacionales) representan solo una parte del ecosistema específico que ellas debían proteger. La identificación de "zonas amortiguadoras" alrededor de tierras valiosas de parques puede resultar útil para aliviar la presión sobre el ecosistema más amplio dentro del cual se encuentran los parques.

Con frecuencia se omite de este esfuerzo de identificación una evaluación comparable de la propiedad de la tierra y de sus usos. Tales factores socioeconómicos pueden ser especialmente importantes donde los derechos de uso se han desarrollado a lo largo de un período histórico extenso, reflejando modelos culturales más amplios de las relaciones entre dueño y arrendatario, la expansión en la frontera activada por el crecimiento demográfico que comprime la capacidad de las áreas urbanas para mantenerse y los esquemas de desarrollo auspiciados públicamente para producir cosechas destinadas a la exportación que permitan obtener dinero en efectivo rápidamente. Además, el uso de la tierra, el rango y los modelos de asentamiento de los pueblos autóctonos quizás necesiten mayor cantidad de documentación, especialmente en las áreas consideradas como "zonas amortiguadoras" situadas alrededor de las áreas de conservación.

La combinación de la información en lo referente a los valores del hábitat y la tenencia de tierra puede hacerse con la ayuda de un Sistema de Información Geográfico por Computadora (GIS), que organiza la información en "capas" que pueden examinarse por separado o en combinación con variables contenidas en otras capas. La tecnología es conocida por las organizaciones de conservación de los países panamericanos, pero quizás en el momento no se esté aplicando a los problemas de trazados geográficos institucionales.

Una transferencia de tecnología de este tipo aportaría por lo menos dos beneficios. Primero, las posibles áreas de hábitat a escogerse podrían evaluarse sistemáticamente y se darían rangos en cuanto a prioridad en base a una combinación de factores socioeconómicos, así como también de factores biofísicos. Potencialmente podría reforzarse la base de planificación para lograr esfuerzos de conservación sostenidos y a largo plazo. Tal como podrá observarse en el caso costarricense descrito en este documento, para que los gobiernos puedan aprovechar al máximo los éxitos iniciales, es necesario que cuenten con una serie de proyectos ya preparados para que pueda recomendarse su ejecución. El segundo beneficio resultante de la promoción de la transferencia de la tecnología del sistema de información geográfica sería el uso más amplio de esa capacidad analítica mejorada dentro del país, en organizaciones que han aumentado la sensibilidad ante las inquietudes locales, lo que contribuiría a limitar el uso de expertos de fuera del país. En consecuencia, los proyectos que se recomendarían para ejecución probablemente reflejarían con precisión los intereses y las prioridades locales, componentes clave para el éxito dentro del marco institucional apropiado.

Resúmenes de tres casos: Bolivia, Costa Rica y Ecuador

1. Bolivia

La primera conversión de deuda por naturaleza fue esencialmente el convenio entre una organización ambiental y el gobierno boliviano para ayudar a financiar y garantizar que más de 1,5 millones de hectáreas (4 millones de acres) de bosques y praderas tropicales sean administradas de conformidad con las normas de conservación.

Conservation International, una organización sin fines de lucro con sede en Washington, compró $650.000 de la deuda contraída por Bolivia con la banca comercial, utilizando al Citibank como su agente en los mercados financieros secundarios. La deuda se adquirió con un descuento de aproximadamente 85% -- es decir, 15 centavos por cada US$. Una subvención de $100.000 de la fundación Frank Weden se destinó a financiar esta transacción.

De conformidad con los términos del convenio de 1987, quedaría cancelada la obligación de Bolivia de pagar la deuda de $650.000. A cambio de ello, el gobierno boliviano acordó dar la máxima protección legal a la Reserva de Biosfera del Beni, en el nordeste del país, y aumentar a 1,2 millones hectáreas (3,7 millones de acres) las áreas protegidas cercanas a la reserva. El gobierno también acordó suministrar un fondo operativo en moneda local por un valor de $250.000 para administrar la Reserva del Beni y sus zonas amortiguadoras adicionales para usos sostenibles.

Los bosques y las praderas de la Reserva de Biosfera del Beni aloja 13 especies en peligro de extinción y a 500 especies de aves. La Reserva es también el hogar de los indios chimanes. Las áreas recién protegidas se administrarán para emplear el bosque de una manera sostenible, protegiendo los suelos y las cuencas hidrográficas a la vez que se protegen los medios de subsistencia tradicional de los chimanes.

La organización "Conservation International" se ha convertido en un asesor científico y técnico del gobierno boliviano en cuanto a conservación y gestión de los recursos naturales. Para llevar a cabo los programas de conservación, la organización acordó designar como su representante a una comisión nacional recién creada, sujeta a la aprobación del Ministerio de Agricultura.

La comisión nacional incluye a funcionarios locales de la región de Beni, a científicos universitarios del Instituto de Ecología y de la Estación Biológica del Beni, así como un representante de una coalición ambiental, la Liga de Defensa del Medio Ambiente (LIDEMA). Esta comisión es responsable de llevar a cabo los programas ambientales.

El fondo operativo de $250.000 en moneda local se creó con el equivalente a US$100.000 en pesos del gobierno boliviano y con $150.000 en pesos aportados por la Agencia para el Desarrollo

Internacional (AID) de sus fondos en moneda local PL480.[5]

La administración del fondo operativo corresponde al Ministerio de Agricultura y a la institución no gubernamental boliviana.

De la experiencia de la conversión de deuda efectuada en Bolivia puede concluirse que las percepciones erróneas pueden causar una reacción nacionalista. Ningún grupo extranjero obtuvo control de la tierra en Bolivia, aunque algunos artículos de prensa dieron esta impresión. En efecto, la tierra ya era propiedad del gobierno boliviano y sigue estando bajo control del gobierno. El intercambio de deuda por naturaleza está dirigido a respaldar un esfuerzo de protección y de gestión de los recursos más coherente y a largo plazo.

2. Costa Rica

Costa Rica creó un sistema de conversión de la deuda que permite mancomunar los fondos de los donantes de deuda para financiar una multitud de programas. Inicialmente, el gobierno estableció en 1987 un programa de conversiones de deuda por naturaleza de US$5,4 millones para crear un Fondo de Conservación de Recursos Naturales utilizando bonos emitidos especialmente para apoyar el esfuerzo. La cantidad máxima de US$5,4 fijada originalmente fue sobrepasada rápidamente por las ofertas de los donantes, y, según se ha indicado, la cantidad total de fondos para conservación generados hasta agosto de 1990 fue el equivalente a más de 40 millones de colones.

En este esfuerzo participan muchos donantes. Los fondos para comprar deuda provinieron de Nature Conservancy, del World Wildlife Fund (incluyendo sus organizaciones nacionales de Estados Unidos, Canadá y Europa), de la Asociación Ecológica La Pacífica, del Pew Charitable Trust, del MacArthur Foundation, de JS Noyes Foundation, de la Sociedad Sueca para la Conservación de la Naturaleza, y de la Fundación W. Alton Jones, de la Organización para los Estudios Tropicales, y de Conservation International.

Otro donante que vale la pena mencionar es el Grupo Financiero Fleet Norster, que, en 1988 se convirtió en la primera institución financiera que realmente donó la deuda para una conversión de deuda por naturaleza. Mediante Nature Conservancy, donó más de US$250.000 (cantidad nominal) en títulos de la deuda costarricense destinados a los esfuerzos de conservación en el parque nacional Braulio Carrillo. Esta donación ocurrió luego de un dictamen del Servicio de Impuestos Internos de Estados Unidos que estipulaba que los bancos pueden donar deuda para fines caritativos y deducir de sus obligaciones fiscales una cantidad igual al valor de mercado de la cantidad en moneda local

[5] "PL-480 funds" (Fondos 480-PL) es un término utilizado para uno de los dos tipos de deuda contraída con el gobierno de Estados Unidos. El desembolso de fondos puede autorizarse para operaciones internacionales de crédito según la Ley de Ayuda Externa de 1961 (enmendada), y también se autoriza el uso de los fondos para el crédito internacional por la ley pública 83-480, Ley de Comercio Agrícola, Desarrollo y Asistencia de 1954".

recibida en una conversión, siendo el balance del valor nominal aceptable como una pérdida de la empresa.[6]

La deuda de Costa Rica se estaba vendiendo en aproximadamente 17 centavos el dólar en 1988, lo que representaba para los compradores un descuento de más del 80 por ciento. A cambio de sus pagarés, Costa Rica ofrecía bonos a 75 por ciento de su valor nominal. Estos bonos, emitidos en colones, con plazos de vencimiento de hasta seis años, daban un interés promedio de 25 por ciento. Solo los pagos de interés están disponibles de inmediato para financiar proyectos ambientales porque no pueden venderse los bonos. El capital de los bonos solo se paga en el segundo año. Sin embargo, cuando se necesita más dinero en efectivo, los bonos pueden emplearse como garantía colateral para los préstamos. Las conversiones son administradas por el Banco Cooperativo Costarricense (BANCOOP) en nombre de la Fundación de Parques Nacionales y la Fundación Neotrópica, que canaliza fondos para proyectos acordados por los donantes y el Ministerio de Recursos Naturales, Energía y Minas.

Los fondos de conservación recién generados se están utilizando para ampliar, administrar y proteger muchos de los parques de Costa Rica, creando la infraestructura, tanto para el turismo como para la ciencia. Los costos fijos de los parques, ya establecidos por ley, absorben hasta 90 por ciento de los fondos existentes presupuestados por el gobierno. No se ha contado con más dinero público para nuevos programas, para reforestación o para efectuar otros trabajos de educación de la población local en los temas ambientales y así impedir la deforestación.

Un esfuerzo importante que está beneficiándose del mecanismo de conversiones de deuda por naturaleza es el proyecto del Parque Nacional Guanacaste. La finalización de este parque requiere compra de tierras en un área ubicada entre los parques y las reservas existentes. Cuando el parque esté terminado, el gobierno costarricense obtendrá el título de propiedad del parque. El nuevo parque será lo suficientemente grande para permitir la restauración de un ecosistema que ha desaparecido en otros sitios -- el bosque tropical seco. La Universidad de Pennsylvania ha aportado una gran cantidad de asistencia técnica.

El trabajo con la comunidad local es parte esencial del plan de Costa Rica para que las personas puedan percibir cuánto valor tiene el parque para ellos. La organización no gubernamental local toma parte activa en el diseño y la administración de los programas financiados por la conversión de deuda por naturaleza.

---

[6]Este dictamen, conocido formalmente como Dictamen de Ingresos 87-124, fue emitido en noviembre de 1987 por el Departamento de Tesorería de los Estados Unidos. Su propósito principal es colocar al prestamista en la misma posición, independientemente de si la deuda se dona para fines caritativos, o se vende y luego las ganancias se donan a las obras caritativas.

3. Ecuador

La Fundación Natura, un importante grupo privado de conservación de Ecuador, llegó a un acuerdo con la junta monetaria del gobierno en 1987, lo que permitió a la Fundación intercambiar hasta US$10 millones de la deuda por bonos en moneda local. Estos bonos iban a emplearse para financiar el amplio rango de actividades de conservación llevadas a cabo por la Fundación y pãra mejorar los parques nacionales de Ecuador, entre ellos la reserva ecológica Cayambe-Coca, el Parque Nacional Podocarpus, la Reserva Natural Pasochoa, y el Parque Nacional Galápagos.

La Fundación Natura buscó donaciones locales e internacionales para adquirir esta deuda. El World Wildlife Fund acordó comprar el primer millón de sucres por apenas un poco más de 35 centavos el dólar; así con US$354.000 se compró 1.000.000 de sucres de los préstamos de la banca comercial a Ecuador.

El Gobierno emitió bonos por la cantidad total del pagaré de la deuda, 100 por ciento del valor nominal, y fueron convertidos a sucres a la tasa de cambio oficial. (La tasa de cambio oficial es sustancialmente inferior a la tasa flotante.) El interés de estos bonos, llamados Bonos de Estabilización Monetaria, está vinculado a las tasas del mercado -- actualmente es de cerca de 33 por ciento de interés anual -- y vencimiento a más de 9 años. El capital se convertirá en una donación para Fundación Natura, mientras que los intereses pagan sus programas en curso.

La Fundación Natura no es el único beneficiario, porque servirá para canalizar los fondos destinados a apoyar los esfuerzos de conservación que incluyen a otros grupos no gubernamentales y a programas públicos. Los programas apoyados por estos bonos incluirán planes de gestión de las áreas protegidas, desarrollo de infraestructura de parques, adquisición de las reservas naturales pequeñas y adiestramiento del personal de los parque, además de actividades de educación ambiental más amplias.

En territorio ecuatoriano se encuentra una amplia variedad de hábitats valiosos, desde las selvas tropicales amazónicas hasta los altiplanos andinos; y desde la sabana costera hasta los manglares de la costa del Pacífico. Sus Islas Galápagos, albergan a plantas y animales que no se encuentran en ningún otro lugar del planeta. Hasta la fecha, solo el Parque de las Islas Galápagos ha producido ingresos netos. Los ingresos del turismo exclusivamente en las Islas Galápagos han aportado el presupuesto para los otros 12 parques ecuatorianos, que cubren más del 10 por ciento de la superficie terrestre del país. Si bien el programa de 10 millones de sucres quizás no influya sustancialmente en la deuda del país, los interese generados por los bonos en el primer año casi ha duplicado el presupuesto del gobierno nacional para los parques nacionales.

Conclusiones y recomendaciones

La característica que constituye el atractivo principal de las conversiones de deuda por naturaleza es que mediante ellas no sale dinero del país deudor. En efecto, lo que ocurre no es tanto un intercambio sino una reasignación del gasto gubernamental: se pasa del pago de intereses a prestamistas extranjeros al manejo de los recursos naturales dentro del país mismo. Se recomienda que se estudien las oportunidades para dar incentivos a las instituciones de crédito de los países acreedores para apoyar este mecanismo de conversiones. Ayudarán a lograrlo los nuevos incentivos fiscales y las prácticas de contabilidad revisadas. A mi parecer, lo más importante del reto se halla en dar a las instituciones de crédito garantías de que los fondos se emplearán adecuadamente para lograr metas declaradas de conservación. Esto es tanto un problema de percepción y de comunicación transcultural, como un problema técnico. Una contribución importante a la superación de este problema puede ser el mejoramiento de la capacidad de las instituciones locales para controlar las iniciativas de conservación.

La participación de las organizaciones no gubernamentales es esencial, y los gobiernos deben procurar activamente dar incentivos para tal participación. Existen oportunidades para la intervención federal directa, como la legislación propuesta por el Congreso y la Iniciativa de las Américas que el Presidente Bush presentó en junio de 1990.[7] Los gobiernos también pueden desempeñar una función indirecta. Se recomienda promover la transferencia de tecnologías como una forma de incentivar la participación de las ONG. En particular, una transferencia de tecnologías puede dar lugar a una mejor recopilación y gestión de la información sobre el valor del hábitat y la tenencia de tierra, y una mejor comprensión de su interacción. De esta forma, los programas propuestos de conservación pueden estar más afianzados en el conocimiento actual del entorno social, económico y biofísico de los programas, y las organizaciones que generan tal conocimiento tendrán más poder sencillamente por hacerlo.

La Iniciativa se propone favorecer los logros económicos para América Latina y países del Caribe y para Estados Unidos. Para optar a cualquier forma de condonación de la deuda y de reestructuración es necesario cumplir con ciertos criterios económicos. Estos criterios incluyen: (1) un programa de reforma económica del FMI/Banco Mundial; (2) diversas reformas en el comercio y las inversiones; y (3) convenios satisfactorios con la banca comercial.

Si reúne estos requisitos, un país podría optar por recibir condonación y reestructuración parcial de su deuda, y por último, establecer fondos fiduciarios para el ambiente. Debido a estas condiciones estrictas, sin embargo, muchos países todavía no reunirían los requisitos (por ejemplo, Brasil y Perú). En el caso de un país apto, el gobierno de Estados Unidos condonaría parte de la deuda y permitiría que los intereses sobre la parte restante se pagara en un fondo fiduciario en moneda local a una tasa de interés concesionaria. Luego, el dinero del fondo fiduciario financiaría programas ambientales.

---

[7] La Empresa de la Iniciativa de las Américas fue anunciada por el Gobierno del Presidente Bush en junio de 1990. La Iniciativa consta de tres partes principales, a saber: (1) Comercio -- establecer normas de comercio libres y justas tales como reducciones recíprocas en las barreras al comercio; (2) Inversión -- introducir reformas importantes en las inversiones que pueden incluir privatización de empresas estatales y la liberalización de los regímenes de inversión. La Iniciativa requiere que el Banco Interamericano de Desarrollo desempeñe una función significativa para efectuar estas reformas; y (3) Deuda -- permitir la condonación parcial de la deuda soberana contraída con el Gobierno de Estados Unidos a los países que puedan optar a ello, tanto de América Latina como del Caribe. Esto no se aplica a la deuda contraída con la banca comercial.

La Participacion de las Comunidades en la Proteccion del Habitat Natural de los Primates Neotropicales y la Politica de Recursos Naturales en Bolivia

Hernán Zeballos H.

Subsecretario de Desarrollo Agropecuario
Ministerio de Asuntos Campesinos y Agropecuarios (MACA)
LaPaz, Bolivia

1. IMPORTANCIA DE LOS RECURSOS NATURALES RENOVABLES

i. En la conservación de los eco-sistemas
ii. En el manejo adecuado de los recursos
iii. En los beneficios específicos de cada recurso individual
iv. En la integración del hombre como recurso natural

2. IMPORTANCIA DE LOS PRIMATES NO HUMANOS

En lo ecológico:

i. Por la dispersión de frutos y propagación de semillas
ii. Por la relación entre la floración y fructificación
iii. Por ayudar a definir la presencia de asociaciones vegetales
iv. Por la influencia e interrelación entre ciertas especies animales y su nicho ecológico
v. Por influir en el bosque con la dinámica del ecosistema

En lo científico:

i. Por ser el hombre un primate hay estrecha relación filogenética entre el y los primates no humanos
ii. Por que los primates no humanos, se convierten en excelentes modelos para investigación biomédica
iii. Por ayudar a comprender al hombre en su anatomía fisiología y aún en su comportamiento social

En lo cultural:

i. Por que al ser los primates no humanos tan similares al hombre se constituyen en modelos didácticos para la enseñanza de las ciencias biológicas
ii. Por su fin lúdico, como mascotas y especies de zoológico donde el hombre encuentra complementariedad en su comprensión de una vida a fin a la suya

En lo social:

i. Por su papel epidemiológico como portadores de enfermedades transmisibles al hombre como malaria, fiebre amarilla y otras
ii. Por su importancia económica para los pobladores de la selva primaria como parte de su sustento, el cual debería regularse en una etapa posterior como posible fuente de ingresos si se hace un manejo adecuado de los mismos en zoo-criaderos o reservas de cría

3. EL ROL DE LA PRESERVACION DE LAS ESPECIES PARA EL BIENESTAR DE LA HUMANIDAD

i. En primer término, tiene un papel ético por la relación del hombre y el medio ambiente, ya que el hombre mismo es fruto de la evolución biológica

ii. Para incrementar el relativo escaso conocimiento del hombre sobre la diversidad biológica del planeta y como foco para preservar la máxima bio-diversidad que podría reportarle al hombre beneficios insospechados en un futuro próximo,

iii. Para mantener el equilibrio ecológico, ya que cumplen un rol específico desde las minúsculas bacterias, hasta las especies de gran tamaño como las ballenas y los elefantes, ejerciendo influencia sobre el resto de las especies vegetales y animales,

iv. Constituyéndose en parte del patrimonio natural de los países y una riqueza intangible, pero posiblemente en el futuro si avanzamos en su conocimiento, en una real fuente económica,

v. Para cumplir con el valor ético de las actuales generaciones de dejar a las futuras un legado similar al que recibieron.

4. LA IMPORTANCIA DE LA PRESERVACION DE LOS PRIMATES NO HUMANOS EN EL CASO BOLIVIANO

Si en el contexto universal señalado antes, se hace necesario preservar a ese medio-hermano del hombre, los primates no humanos, con mayor razón en Bolivia por.

i. La diversidad ecológica

Bolivia, utilizando de la clasificación de zonas de vida de Holdridge, cuenta con: 3 regiones fisiográfica principales y 27 zonas de vida (ver mapa 1 y tabla 1).

ii. Contar con 557.000 Km2 de bosques (algo más del 50% del territorio nacional) en 2 grandes regiones (ver cuadro No.2).

iii. Que las extensas regiones de bosques amazónicos y no amazónicos representan una riqueza potencial en flora y fauna silvestres. En términos de fauna se conoce que existen aproximadamente,

	1.270...............especies de aves
	300............especies de mamíferos
más de	800...............especies de peces
alrededor de	100...............especies de anfibíos

En términos de flora se estima que existen más de 20,000 a 22.000 especies[1]

En esa magnífica diversidad se estima que existen aproximadamente,

18...............especies de primates no humanos y,
22...............subespecies[2]

iv. El uso irracional que se hace de la abundante y diversa fauna primatológica de Bolivia y por que esta se encuentra deficientemente conocida.

El uso irracional se presenta tanto en las actividades que tienen una finalidad en la alimentación como en la caza para la obtención de mascotas; y para las actividades orientadas hacia la exportación de animales, hacia los países desarrollados.

La fauna primatológica es también afectada por los procesos de modificación de los habitats, modificaciones que alcanzan grandes proporciones y se deben principalmente a la falta de una planificación en el desarrollo y colonización de las áreas silvestres.

v. El poco conocimiento que se tiene de estas especies según anota Heltne[3] , los estudios en terrenos sobre poblaciones de primates no humanos en Bolivia, son escasos y largamente distanciados en el tiempo.

Los primeros estudios fueron realizados en La Paz por un Museo Alemán que envió una misión para recolectar animales. Cincuenta y cinco (55) años más tarde una misión auspiciada por la OPS/OMS[4] realizo investigaciones en los departamentos de Santa Cruz, Beni, Pando y Norte de La Paz.

---

[1]Beck, Stefan, Director Herbario Nacional, comunicación personal.

[2]Anderson (1985) Lista preliminar de mamíferos bolivianos. Cuadernos Academia Nacional de Ciencias de Bolivia. VI Ciencias de la Naturaleza Zoología 3, La Paz, 16p.

[3]Heltne, Paul (1975). "Investigación en el terreno sobre poblaciones de Primates no Humanos en Bolivia", OPS-OMS. 25p.

[4]Anderson (1985). *op. cit.*

Mas recientemente Isawa (1979-81), Yonida (1981-84), Masataka (1981), Minisawa (1981-84), Poola (1979-82) y Garcia (1986-88) realizaron trabajos en el terreno, los cuatro primeros en los departamentos de Beni y Santa Cruz.

5. LA POLITICA BOLIVIANA EN TORNO A LOS RECURSOS NATURALES Y EL MEDIO AMBIENTE

Por su ubicación en el Centro del Continente Sudamericano, Bolivia comparte los diferentes ecosistemas del mismo, siendo un país de tradición e imagen andina, en realidad la mayor parte de su territorio está constituida por tierras bajas tropicales, pertenecientes a las Cuencas de los ríos Amazonas y el Plata. A su diversidad geográfica, se añade su diversidad cultural.

Un complejo proceso histórico ha conservado la presencia de numerosos pueblos indígenas originarios, tanto en las regiones andinas como en las Amazónicas del país.

El deterioro creciente de los ecosistemas provocado por la aplicación de inadecuados modelos de desarrollo en los cuales no se tuvo en cuenta el uso sostenido de los recursos, trajo como consecuencia el actual deterioro del medio ambiente y la explotación irracional de los recursos naturales renovables. Frente a ésta situación el gobierno de Bolivia se encuentra ejecutando una política de desarrollo donde los factores ambientales tanto de preservación como de uso sostenido, sean incorporados al desarrollo económico y social del país.

Se ha decretado la vigencia de una "Pausa Ecológica Histórica de Cinco Años" la cual prohíbe: nuevas concesiones de aprovechamiento forestal, caza y acoso de animales silvestres y regula las acciones para un mejor control y conservación del medio ambiente.

i. La Pausa Ecologica

Se funda en la necesidad de establecer el espacio de tiempo indispensable para permitir un reordenamiento de todos aquellos procesos que, por su impacto sobre la naturaleza, hacen peligrar la sustentabilidad de la base material de la vida humana y del patrimonio natural boliviano.

Mas aún, consideramos que los recursos naturales de nuestro país, no pueden tener más el forzoso destino de su destrucción, como único camino para elevar la calidad de vida del hombre boliviano. Nuestro medio ambiente constituye un potencial productivo al que le asignamos la mayor importancia, en beneficio colectivo, de nuestro país y del mundo.

La pausa ecologica historica marca un hito en la dinámica de los procesos de desarrollo y su relación con la conservación en Bolivia.

Esta voluntad constituye un compromiso y un desafío inéditos ante el pueblo boliviano y su historia. Pero supone también una puesta al día de Bolivia con la preocupación y las acciones que la comunidad internacional manifiestan sobre este tema: comporta un compromiso contraído con ella.

Aceptamos el criterio de la corresponsabilidad internacional frente al deterioro global del medio ambiente. Los modelos de desarrollo mundiales muestran desajustes. El progreso y la teconolgía han traído también degradación ambiental y, lo que es peor, no han resuelto el problema de la marginalidad y la pobreza en países como los nuestros. Pero entendemos también que, sobre la base de una responsabilidad compartida, pueden constituirse relaciones internacionales más creativas, más justas, que nos permitan encarar el futuro con esperanza.

ii Administracion de los Recursos Naturales

La administración de los recursos naturales y medio ambiente, se efectúa a través de la Subsecretaria de Recursos Naturales Renovable y Medio Ambiente, cuyo objetivo y finalidad es la de normas y fiscalizar las políticas y estrategias nacionales del medio ambiente y los recursos naturales, a través de una planificación y programación de actividades de ejecución, control y evaluación de los recursos hídricos, forestales, suelos, aire, flora, fauna e hidrobiológicos. Asimismo, su función abarca la aplicación de normas técnico-operativas de acciones de control del medio ambiente y áreas protegida en áreas rurales y urbanas.

Unidades que conforman la Subsecretaría:

- Dirección Nacional del Medio Ambiente
- Dirección Nacional de Recursos Integrados
- Dirección Nacional de Recursos Pesqueros
- Centro de Desarrollo Forestal
- Centro de Desarrollo Pesquero
- Instituto indigesnista Boliviano

Campos de Actividad:

Los campos de actividad de la Subsecretaría son los de investigación, planificación, control, normas y fiscalización de los recursos naturales y medio ambiente, comprende acciones de control y protección de los bosques, depredación de la fauna silvestre, conservación de parques nacionales, control de la contaminación ambiental, preservación de los recursos suelo, aguas, fauna, flora y aire.

Asimismo formula y dirige las políticas para las poblaciones indígenas y tribales del país.

iii Ley del Medio Ambiente y Recursos Naturales

La aprobación de la Ley del Medio Ambiente y Recursos Naturales constituye uno de los principales objetivos del gobierno, puesto que su contenido legislará en forma integral a todos los sectores involucrados en la problemática ambiental (flora, fauna, aire, agua, suelo).

Será un instrumento moderno que permitirá reestructurar todo el aparato estatal relacionado al medio ambiente para lograr de esta forma un aprovechamiento sostenido de los recursos y preservar la naturaleza para las futuras generaciones.

La ley del Medio Ambiente se encuentra en este momento en etapa de revisión por una comisión especial de alto nivel, y se considera que la misma será promulgada dentro de un año y medio.

6. LA POLITICA BOLIVIANA EN TORNO A LOS RECURSOS NATURALES RENOVABLES

i. Administración de los RRNA

El Centro de Desarrollo Forestal (CDF) constituye el organismo rector de la política en la materia.

Sus funciones se refieren a: la protección, el manejo, aprovechamiento, transporte y comercialización de animales de fauna silvestre y de los productos que puedan abstenerse de aquellos, así como también la protección de las especies amenazadas de extinción, la conservación del hábitat para la fauna y flora, establecer parques nacionales y otras actividades que tiendan a la conservación, fomento y aprovechamiento racional de los recursos naturales.

ii El proyecto de- Ley del Medio Ambiente, se encuentra en proceso de aprobación por el Congreso nacional.

iii. Estado de protección de los primates.

Indirectamente se encuentran protegidos por dos disposiciones internacionales, la convención de CITES y el libro rojo de las especies (red databook) auspiciado por CVN.

Por los anterior es de alto interés para el gobierno de Bolivia, Ministro de Asuntos Campesinos y Agropecuarios - MACA; el establecer programas y proyecto específicos como el convenio, nacional boliviano de primatología a suscribirse con la OPS/OMS, que permita mejorar el marco legal y la conservación y producción económica de los recursos naturales.

7. PARTICIPACION DE LAS COMUNIDADES NATIVAS EN LA CONSERVACION DEL HABITAT NATURAL DE LOS PRIMATES NEOTROPICALES

Bolivia ha sido recientemente conmovida por la denominada marcha indígena "por la dignidad y territorio"

Alrededor de 800 indígenas realizaron una marcha de 620 kilómetros desde la catedral de

Trinidad, en el Beni, hasta la catedral en la ciudad de La Paz.

La realización de esta marcha, que traduce, las presiones ante los grupos sociales de mayor poder económico (ganaderos y empresas forestales y las etnias y movimientos migratorios de los campesinos de la zonas alto-andinas hacia el subtrópico, obligó al gobierno central, a emitir disposiciones legales en favor de los grupos étnicos del oriente boliviano.

Cuatro Decretos Supremos, del 22609 al 22612 "declaran la necesidad nacional y social, de reconocimiento, asignación y tenencia de áreas territoriales en favor de grupos selvícolas y comunidades indígenas del Oriente y la Amazonia.

Reconocen también estos decretos "los convenios internacionales que contemplan el derecho de las poblaciones indígenas a los recursos naturales existentes en sus tierras y la protección especial de los mismos".

Es en este nuevo marco legal en donde se abren posibilidades interesantes en cuanto menos dos direcciones. La primera, para aprovechar la habilidad de los grupos selvícolas de convivir en los bosques y con los bosques, y la segunda, para establecer proyectos que aprovechen, sin romper el equilibrio de los ecosistemas, los recursos de flora y fauna para comercio regular, particularmente científico, generando así ingresos para esas comunidades.

Los proyectos vinculados a estas etnias pueden servir por tanto a los siguientes propósitos:

i. La generación de ingresos,

ii. El servir de vínculo entre las comunidades y los poderes estatales para brindar servicios de educación y salud,

iii. Para dar asistencia técnica a nuevas formas de producción agrosilvopastoril,

iv. Para retornar a las comunidades los resultados de las investigaciones médicas, ecológicas y otras vinculadas a los proyectos.

PRIMERA PROPUESTA DE UN PLAN DE ACCION REGIONAL DE PRIMATOLOGIA PARA LAS AMERICAS

Filomeno Encarnación

Hugo Samamé

Centro de Reproduccion y Conservacion de Primaes No Humanos
Projecto Peruano de Primatologia
Instituto Veterinario de Investigaciones Tropicales y de Alturas (IVITA)
Universidad Nacional Mayor de San Marcos
Apartado Postal 621
Iquitos, Peru

I. INTRODUCCION

Los 157 Estados miembros de la OMS, colectivamente, durante la Vigésima Octava y Vigésima Novena Asamblea Mundial de Salud, expresaron su preocupación por el peligro que corren las especies de primates no humanos por causas de la destrucción de sus hábitats naturales. En las resoluciones WHA 28.83 y WHA 29.67, se ordena que el Director General WHO facilite el intercambio de recursos y de tecnología entre los países afectados y con la ayuda de organismos internacionales, suministre el asesoramiento sobre la conservación, crianza y reproducción de las especies.

Los representantes oficiales de Perú, México, Canadá, USA, Brasil, Guatemala, España y Sud Africa, funcionarios de AID de los Estado Unidos, investigadores de las Universidades de USA y Perú, de organizaciones no gubernamentales del Perú que asistieron al "Taller Sobre Crianza Controlada de Primates No Humanos en su Propio Hábitat" realizada en Iquitos, en Noviembre de 1986 firmaron la "Declaración sobre la cooperación técnica, entre los países de las Américas, en la conservación y utilización de los primates no humanos". En el referido documento se recomienda la creación de un Comité Regional de Primatología para las Américas, en base a la experiencia desarrollada por el Proyecto Peruano de Primatología "Manuel Moro Sommo", al mismo tiempo se solicita la colaboración de la OPS, a fin de elaborar las estrategias y planes de acción regional promoviendo la cooperación interpaíses que posibiliten el desarrollo y fortalecimiento de los programas nacionales para la conservación, reproducción y utilización de los primates neotropicales.

En la RIMSA V celebrada del 27 al 30 de Abril de 1987, la delegación del Perú puso en consideración la "Declaración sobre la cooperación técnica, entre los países de las Américas, en la conservación y utilización de los primates no humanos", de modo que en la Resolución VIII se recomienda para que la OPS, en actuación de secretaría ex-oficio, coordine el establecimiento del Comité Regional de Primatología.

IMPORTANCIA DEL PLAN REGIONAL

El Plan Regional constituye una guía y marco de referencia, dentro de la voluntad política de los países de la Región en la que se encuadran los planes nacionales para la conservación y protección de las especies de primates neotropicales así como de su hábitat natural, procurando la concurrencia del esfuerzo regional y facilitanto la definición de acciones y solidarias entre los países miembros.

El Plan Regional proporciona la orientación fundamental para la política de conservación, de organización de procedimientos y de programas cooperativos interpaíses miembros de la OPS. En el ámbito mundial, representa la participación de la región en la instrumentación de estrategias globales de conservación y aprovechamiento de los primates no humanos en su propio hábitat, concordantes con las resoluciones WHA 28.83 y WHA 29.67 de las Asambleas de la OMS.

Considerando cierta homogeneidad y similitud en la ecología y dinámina del sistema de bosque tropical húmedo del neotrópico, las metas y los objetivos para la conservación de los Estados Miembros, son concurrentes acordes con sus particulares situaciones y posibilidades, de modo tal que en esfuerzo conjunto se pueda garantizar la conservación y perpetuación de los primates no humanos.

II. SITUACION DE LOS PRIMATES EN EL NEOTROPICO

La vida de las poblaciones naturales de primates neotropicales, está directamente relacionada con la presencia de los bosques naturales, en cuyo espacio vertical y horizontal cumplen sus funciones de reproducción, alimentación, descanso y otros. Sin embargo, la presencia del bosque en el trópico húmedo en general, depende del grado y nivel de entendimiento de la población humana para conservarla y protegerla.

Grandes extensiones del bosques están siendo modificadas por el uso del suelo bajo patrones ortodoxos y tradicionales, en algunos casos tan acelerados mediante el uso de maquinarias y equipos, que el mismo hombre va inventando, para su aplicación en la agricultura o tala desde las zonas templadas, ocasionando la eliminación o degradación de los mismos.

Debemos tener presente que el bosque húmedo tropical constituye el ecosistema más rico de la tierra y lamentar que abarca sólo una extensión del 8% del total de nuestro planeta. En este bosque viven entre 40 y 50% de las especies del total de la biodiversidad estimado en 5-10 millones. Pero, esta bioma es la menos conocida comparada con las otras, con excepción de los primates no humanos que sí son medianamente conocidos, aún cuando la pérdida del hábitat boscoso es un factor predominante para la disminución de las aquellas especies.

La humanidad vive momentos de fuertes y álgidas discrepancias entre dos concepciones sobre el uso del bosque, por un lado se afirma que la mayor disturbación o alteración conduce al empobrecimiento, y por otro, que la remosión o tala total y permanente favorece su recuperación y optimiza su aprovechamiento. Datos estimados en la década del 70, por dos investigadores, son referidos en 200,000 km^2/año de bosque tropical afectos a la degradación y destrucción (Myers 1980) por acción antrópica, y 76,000 km^2 de bosque eliminados cada año como resultado de la comercialización de maderas (Lanley 1982).

En América latina, los agentes principales de la deforestación, son el auge de la ganadería que ocasiona la tala rasa de unos 20,000 km^2/anuales, las plantaciones de monoculturas de palma aceitera, té, cacao, coca, caña de azúcar, plátanos, y otros, que en gran porcentaje son productibles hasta por unos 50 años.

En futuras décadas, la pérdida o el empobrecimiento afectarán marcadamente la continuación de valiosos trabajos críticos sobre evolución. Entre los servicios ambientales advertimos la pérdida de la capa de nutrientes del suelo, desorganización de los sistemas hidrológicos y fuentes de agua, y otros, que repercuten en los cambios climáticos por las alteraciones en el ciclo del CO_2 de la atmósfera global, de modo que afectan la vida de la fauna primatológica. Los ejemplos están en la parte sur y oriental del Brasil (Rondonia, Pará, Matto Grossoy Acre) por la colonización con ganado; en la costa Atlántica del Brasil por la extracción de madera y la agricultura, en el tramo brasileño de la carretera transamazónica, en los tramos de la carretera marginal de la selva, Alto Mayo y Huallaga Central del Perú. En América Central por la agricultura, ganadería y extracción de maderas; en la cuenca amazónica (Caquetá y Putumayo) de Colombia por la colonización y ganadería; en la costa del Pacifíco y Amazonía del Ecuador por la agricultura y extracción de maderas; en la Amazonía (ceja de selva y selva baja) del Perú por la agricultura de coca y ganadería.

En los bosques de difícil accesibilidad que constituyen hábitats de primates no humanos, las evaluaciones de sus poblaciones aún no fueron iniciadas. Pero la tala efecta drásticamente en algunos países, mientras todavía se perfeccionan los métodos y procedimientos de diagnóstico y muestreo de las poblaciones de primates.

Las poblaciones naturales de primates en el bosque con sus patrones de distribución geográfica, abundancia y vulnerabilidad, están relacionadas con los aspectos de la biología de las especies. Vale decir, que el tamaño corporal está directamente relacionada con el período de gestación y el tamaño del territorio, pero indirectamente con la densidad de población; en otras palabras el gran tamaño presenta mayor opción a la predación por el hombre, así como ofrece mayor porción de alimento.

El avance tecnológico con las maquinarias y equipos según la escala, el nivel y la naturaleza de su aplicación, pueden ayudarnos a comprender las implicancias para la explotación o el aprovechamiento del bosque. La explotación destructiva (tala) de especies de plantas como *Castilloa ulei* (para latex), *Couma macrocarpa* y *Ragala sanguinolenta* (para goma) en la cuenca amazónica afecta el recurso alimenticio de *Chiropotes* y *Lagothrix*. La colecta tradicional (tala) de frutos nativos amazónicos como *Rheedia*, *Spondias*, *Inga*, *Caryocar*, *Parahancornia*, etc. merma las posibilidades de alimento de los *Saimiri*, *Saguinus*, *Cebus* y numerosas aves y mamíferos. El clareo o la limpieza del sotobosque para la explotación en pie de árboles de *Hevea brasiliensis* (jebe), y del bejuco *Paullinia cupana* (choco) o la tala para la preparación de leña o carbón en las cercanías de los villorios, disminuyen la fauna entomológica para la dieta de *Saimiri*, *Cebus*, *Saguinus*, *Callithrix*, *Callicebus*, *Callimico*, en la región amazónica.

También las construcciones de carreteras y diques sin las estructuras y recomendaciones técnicas adecuadas, ocasionan embalses, cuyas inundaciones, matan a la vegetación o bosque. Estos ejemplos son objetivos en las represas con fines hidroeléctricos del TUCURUI en el río Tocantins y las Balbinas en Roraima, Brasil, donde ocurrió la inundación y muerte de extensos bosques habitados por primates y otros animales arborícolas.

El manejo agrícola del bosque progresa desde los bosques montañosos hacia las llanuras, de modo que se incrementa la proporción de bosque secundario que inducirá cambios en la estructura y dinámica poblacional (comunidad) de primates. En Colombia se encuentran poblaciones de *Lagothrix lagothricha* y *Ateles belzebuth* en bosques aislados; y de *Cebus albifrons* en bosques secundarios donde prosperan también *Saguinus leucopus*. En Brasil, Manaus, en 20 Ha. de bosque aislado, sobreviven *Pithecia* y *Saguinus bicolor*, ésta última en condiciones normales ocupando el bosque secundario. Circunstancialmente, estos animales adquieren comportamiento semiterrestre, para atravezar las carreteras y áreas desbrozadas.

La tala selectiva de madera comercial, también afecta el bosque por los procedimientos y mecanismos tradicionales aplicados, tales como las concesiones de áreas sin conocimiento de los componentes naturales, la tala y sistema de transporte por arrastre desvastador hacia las masas de agua, la apertura de carreteras y trochas carrozables y forestales, son prácticas comunes en la Amazonía para extracción de Caoba (*Swietenia macrophylla*), Cedro (*Cedrela* spp.), Ishpingo (*Amburana cearensis*), Cumalas (familia ministicácea) y otras. Estos árboles son importantes como sustratos, sin embargo, los exudados y gomas de cedro (*Cedrela* sp.) y quillusisa (*Vochysia* sp.) en el Perú son principales componentes de la dieta de *Cebuella pygmaea*, y los frutos de Virola (Miristicácea) es importante para la alimentación de *Cebus apella* y *Alouatta*.

Igualmente las empresas ganaderas y agroindustriales, estatales y privadas, con las denominaciones de polos o focos de desarrollo u ocupación de la Amazonía, promueven la tala de grandes extensiones con reducción total del estrato arbóreo y proliferación del estrato herbáceo debido a la naturaleza del cultivo elegido. Por ejemplo, pasturas en base a leguminosas y gramíneas, arroz en secano o en pozas, maní, maíz, algodón, tabaco, pimienta y otros. Las consecuencias son drásticas para los animales arborícolas por desaparición de escondites y refugios, del sustrato y las fuentes de alimento. El efecto se potencializa porque se eliminan las semillas (diásporas) y toda posibilidad de regeneración. Estos ejemplos son visibles en diferentes partes de la Amazonía de Brasil, Perú, y Colombia.

La caza de primates no humanos en el neotrópico afecta principalmente a las especies de tamaño grande como *Ateles*, *Lagothrix*, *Alouatta*, *Cacajao* y otros, como fuentes de proteína animal de los pobladores nativos o colonos. Entre Perú y Colombia es práctica común el uso de *Lagothrix* como carnada para cazar felinos (*Felis pardalis*) con piel de buena cotización. El uso con fines ornamentales de pieles, patas, cráneos, es común en la amazonía para preparación de cuadros de pared, llaveros, collares, etc. de especies de tamaño pequeño principalmente de *Saguinus*, *Saimiri*, *Callicebus* y otros. La cola de *Pithecia* tiene mucha acogida como plumero.

La caza de los padres o adultos para obtener animales mascotas toma mucho interés hacia los *Saguinus*, *Cebuella*, *Callithrix*, *Leontopithecus*, *Callimico*, y *Callicebus*; mientras que otras especies como *Saimiri* y *Cebus* son cazados como medida sanitaria por convertirse en invasores de cultivos de maíz, caupí, plátanos y frutales diversos.

La captura para el comercio como mascotas o uso en laboratorios en el presente siglo, sigue ocupando un rol preponderante que repercuten en las poblaciones de primates. Sin embargo, desde Julio de 1975, los gobiernos de 80 países, miembros de IUCN, establecieron pautas y acuerdos para controlar este comercio, con la firma de la convención de Washington - CITES, a la vez que protege las especies en vías de extinción. En el neotrópico, la demanda comercial principalmente desde la Amazonía data sus inicios en la década de 1940, teniendo a los Estados Unidos de Norteamérica el primer país receptor o importador. Algunas cifras, por ejemplo desde el Perú durante 11 años (1964-1974) fue importado casi 340 mil ejemplares vivos, cifra que ha disminuido notablemente, desde 1976 hasta 1989, por las medidas de control del Gobierno del Perú, cumpliendo las recomendaciones CITES y eliminando las causas de la mortalidad ocurridas desde la captura hasta su destino final. Estos resultados se deben a la activa participación, coordinación y asesoría de la OPS/OMS.

Los primates neotropicales habitan en los territorios de los países de América Central y del Sur y en algunas de las Islas Caribe, con excepción de Uruguay y Chile, desde el Sur de México a través de América Central hasta el extremo Sur de Brasil y el Norte de Argentina. La distribución de las especies responde al concepto de las regiones biogeográficas para el neotrópico.
Según Mittermeier (1987), se conocen 16 géneros con unas 65 especies, éstas últimas fueron incrementadas hasta la actualidad con los estudios publicados por Ph. Hershkovitz del Museo de historia Natural de Chicago, M. Ayres del Museo Emilio Goeldi de Belem y otros. Se considera dos géneros, más de 12 especies y un total de 30 sub-especies en situación amenazada y algunas de ellas en verdadera vía de extinción.

Los científicos y autoridades de países con territorios que mantienen poblaciones naturales de primates, realizan esfuerzos nacionales o de coordinación internacional, para captar la atención financiera que implique la protección de las especies mediante los sistemas de unidades de conservación.

En la Amazonía ocurre la mayor diversidad específica, con 12 géneros y más de 40 especies y en el bosque Atlántico del Brasil, con seis géneros y 15 especies; ocupando hábitat preferenciales por debajo de 1000 m.s.n.m. Solamente *Lagothrix flavicauda*, especie endémica del Perú, habita arriba de los 1000 m.s.n.m. entre la línea que divide la región andina con la amazónica. En el endemismo de las especies, radica la importancia de los bosques amazónicos y costa atlántica brasileña, donde las instituciones nacionales con el aporte de organizaciones extranjeras e internacionales, hacen los esfuerzos para evaluar las poblaciones y establecer Unidades de Conservación.

Países como Brasil con más de 51 especies de primates y unas 35 especies endémicas, Perú con más de 28 especies y 11 especies endémicas; y Colombia con más de 27 especies y 7 especies endémicas, han tomado liderazco para la protección de su fauna primatológica y conservación de sus hábitat en los Parques Nacionales, Reservas Comunales, Areas en estudio, y otras.

En la costa atlántica brasileña, desde 1979, se trabaja prioritariamente para salvaguardar los géneros *Brachyteles* y *Leontopithecus*, consideradas amenazadas por la destrucción del hábitat. Los resultados en cautiverio y en hábitat natural, están contribuyendo para delinear ciertos patrones para la recuperación de los bosques y la protección del bosque remanente que además alberga a *Cebus apella xanthosternos*, *Callicebus personatus melanochis*, *Callithrix kuhli* y *Alouatta fusca*. Las acciones fueron asumidas por la Fundación Brasileña para la Conservación de la Naturaleza (FBCN), Universidad Federal de Minas Gerais, y el Centro de Primates de Rio de Janeiro con el apoyo de la WWF principalmente.

III. OBJETIVO GENERAL

Salvaguardar la permanente existencia de las poblaciones silvestres de primates no humanos en su hábitat natural, orientado a compartir los beneficios y mejorar la condición de vida humana en aspectos de salud, educación, agricultura y otros.

IV. ESTRATEGIAS/ACCIONES

1. LA DECLARACION DE AREAS NATURALES PARA LA PROTECCION DE LA FAUNA SILVESTRE EN ESPECIAL PARA LA CONSERVACION DE PRIMATES NO HUMANOS.

1.1 Inventariar y evaluar los ecosistemas con poblaciones nativas.

1.2 Elaborar mapas de distribución geográfica nacionales y regionales.

1.3 Seleccionar y adoptar areas protegidas para que todas las especies estén en salvaguarda.

2. LA PROMOCION Y EL DESARROLLO DE LA COORDINACION INTERSECTORIAL PARA LA EJECUCION CONJUNTA DE PROGRAMAS PARA LA EVALUACION DE HABITAT, CRIANZA, ASI COMO LA VIGILANCIA PERMANENTE DE ESPECIES RARAS O EN VIAS DE EXTINCION.

2.1 Elaborar y ejecutar planes de acción nacionales con la cooperación técnica de la OPS cuando sea requerida.

2.2 Desarrollar y expandir el concepto de política de conservacion y manejo incluyendo el mantenimiento de poblaciones especificas y sus variaciones, con importancia socio-económica actuales y potenciales.

2.3 Desarrollar y adoptar métodos y técnicas del manejo y aprovechamiento sostenido de primates no humanos como una capacidad del bosque para la producción de recursos naturales.

2.4 Elaborar formatos estandares para facilitar el entendimiento técnico-científico entre especialistas y administradores con las agencias financieras y organizaciones ambientales.

3. LA COOPERACION ENTRE PAISES PARA LA CREACION DE AREAS CONJUNTAS PARA LA PROTECCION DE LAS ESPECIES DE PRIMATES NO HUMANOS Y OTROS ANIMALES SILVESTRES.

3.1 Firmar convenios, acuerdos y/o tratados bilaterales entre países limitrofes.

3.2 Formular la cooperación técnica de la OPS en la ejecución de los programas interpaíses de primatología.

3.3 Establecer acuerdos entre países para la transferencia de técnologia

4. LA DISEMINACION DE INFORMACION TECNICA Y CIENTIFICA PARA EL FORTALECIMIENTO DEL DESARROLLO DE PERSONAL Y LOS SISTEMAS DE INFORMACION EN DISCIPLINAS RELACIONADAS A LA PRIMATOLOGIA.

4.1 Establecer e identificar centros de documentación y/o referencia sobre primatología neotropical.

4.2 Capacitar y entrenar a nivel académico y técnico a especialistas.

4.3 Establecer sistemas de información primatológica básica para la conservación de los genes y sus habitat naturales.

4.5 Promover en todos los niveles, la importancia de la conservación del ecosistema y sus recursos genéticos.

4.6 Publicar los resultados de estudios y documentos inéditos, facilitando la difusión e intercambio.

5. LA COOPERACION INTERSECTORIAL NACIONAL E INTERNACIONAL PARA LA FORMULACION Y APLICACION DE PROGRAMAS DE EDUCACION AMBIENTAL.

5.1 Incorporar en los programas y planes de estudios curriculares en todos los niveles a aspectos relacionados a la biología de los primates.

5.2 Promover la participación de la población humana del medio urbano y rural en la conservación y protección de las especies y el ecosistema.

6. EL ESTABLECIMIENTO DE CENTROS DE PRODUCCION EN CAUTIVERIO Y/O SEMICAUTIVERIO DE PRIMATES NO HUMANOS PARA Y CON FINES DE REPOBLACION Y USO SOSTENIBLE EN INVESTIGACIONES BIOMEDICAS.

6.1 Desarrollar planes estandares para investigación, manejo y crianza en cautiverio.

6.2 Desarrollar políticas de manejo humanitario, cuarentena, acondicionamiento y producción de primates no humanos para uso científico en los países nativos y transferencia a terceros países.

6.3 Evaluar los riesgos potenciales de las zoonosis en los primates neotropicales.

6.4 Establecer líneas genéticas de primates.

6.5 Fortalecer ensayos sobre la adaptación en habitat naturales de las especies reproducidas o mantenidas en cautiverio.

7. LA MOVILIZACION DE RECURSOS ECONOMICOS NACIONALES E INTERNACIONALES PARA EL DESARROLLO DE LAS AREAS NATURALES, LOS CENTROS DE REPRODUCCION DE PRIMATES NO HUMANOS Y LOS PLANES Y PROGRAMAS DE CONSERVACION.

7.1 Obtener recursos financieros para capacitación en primatología y ejecución de estudios y/o investigaciones urgentes, críticos o de largo periódo.

7.2 Formular proyectos nacionales, multinacionales de pre-inversión conjuntamente con las agencias financieras bilaterales o multilaterales.

7.3 Establecer un fondo regional para la ejecución de acciones relacionadas a la conservacion de primates neotropicales, asi como fondos nacionales para el mismo fin.

APPENDICES

ANEXOS

First Ordinary Meeting of the Regional Primatology Committee for the Americas (CORP-1)

Primera Reunión del Comité Regional de Primatología para las Américas (CORP-1)

List of Participants
Lista de Participantes

GOVERNMENTS GOBIERNOS	ROLE INTERVENCION	NAME NOMBRE
ARGENTINA ARGENTINA	Representative Representante	Dr. Julio C. Ruiz Centro Argentino de Primates Corrientes
BARBADOS BARBADOS	Representative Representante	Dr. Jean Baulu Programme Leader Barbados Primate Center and Barbados Wildlife Center St. Peter
BOLIVIA BOLIVIA	Representative Representante	Dr. Hernan Zeballos Subsecretario de Asuntos Agropecuarios Ministerio de Asuntos Campesinos y Agropecuarios La Paz
		Dr. Jean Manuel Jesus Angulo Profesor de Microbiologia Universidad Autonoma Gabriel Renee Moreno La Paz
BRAZIL BRASIL	Representative Representante	Dr. Marcos Malacco Presidente Fundacion Nacional de Salud
		Dr. Jordan Paulo Wallauer Secretaria Especial de Medio Ambiente Brasilia
	Observer Observador	Dr. Virgilio Ferreira Fundacion Instituto Qswaldo Cruz Rio de Janeiro
		Dr. Thiago de Mello Presidente Sociedad Latinoamericana de Primatologia
CANADA CANADA	Representative Representante	Dr. James Wong Associate Director of Assessments Canadian Council on Animal Care Ontario
COLOMBIA COLOMBIA	Representative Representante	Dr. Dario Alejandro Cedeño Instituto Nacional de Salud Bogatá

GOVERNMENTS *GOBIERNOS*	*ROLE* *INTERVENCION*	*NAME* *NOMBRE*
COLOMBIA COLOMBIA	Representative Representante	Dr. José Vicente Rodriguez Instituto de Recoursos Naturales (INDERENA) Bogatá
	Observer Observador	Dr. Jaime Umaña Presidente Sociedad Colombiana de Primatología Bogatá
GUATEMALA GUATEMALA	Observer Observador	Dr. Richard Lujan Centro de Investigacione en Enfermedades Tropicales Guatemala
FRANCE FRANCIA	Observer Observador	MS Annie Arsene Universite Louis Pasteur Strasbourg Centre de Primatologie
JAPAN JAPÓN	Observer Observador	Dr. Yasuo Kawanishi Research Primates (JAPAN) Ltd.
		Dr. Kisou Kubota Primate Research Institute
MEXICO MEXICO	Representative Representante	Dr. Alejandro Estrada Estación de Biología Tropica Mexico D.F.
		Dr. Luis Sanchez Torres Gerente Biologícos y Reactivos
PARAGUAY PARAGUAY	Representative Representante	Dra. Maria Teresa Rovira de Muñoz Jefa Proyecto de Primatología Instituto de Investigaciones de Ciencias de Salud Asunción
PERU PERÚ	Representative Representante	Dr. Ricardo Valdivia Rodriguez Director IVITA Lima
		Ing. Manuel Uceda Director General Forestal y Fauna Lima
PUERTO RICO PUERTO RICO	Observer Observador	Dr. Pablo Nieves Caribbean Primate Research Center

GOVERNMENTS *GOBIERNOS*	*ROLE* *INTERVENCION*	*NAME* *NOMBRE*
UNITED STATES OF AMERICA ESTADOS UNIDOS DE AMÉRICA	**Representative** **Representante**	Dr. Roger Anderson Centers for Disease Control Atlanta, Georgia
		Dr. Steve Landry Agency for International Development Washington, D.C.
		Dr. Kirk Miller Agency for International Development Washington, D.C.
		Dr. Stephen Potkay Chief Veterinary Resources Branch Division of Research Services National Institutes of Health Bethesda, Maryland
		Dr. Robert Whitney Director National Center for Research Resources National Institutes of Health Bethesda, Maryland
	Observer Observador	Dr. John Anderson California Primate Research Center Davis, California
		Dr. Anne Baker International Union for Conservation of Nature (IUCN) Species Survival Comission Brookfield Zoo Brookfield, Illinois
		Dr. Gradner Brown University of Washington Seattle, Washington
		Dr. James G. Else Yerkes Regional Primate Research Center Atlanta, Georgia
		Dr. Joe Erwin Diagnon Corporation Department of Primate Ecology Rockville, Maryland

GOVERNMENTS GOBIERNOS	ROLE INTERVENCION	NAME NOMBRE
UNITED STATES OF AMERICA ESTADOS UNIDOS DE AMÉRICA	Observer Observador	Dr. Ronal Hunt New England Primate Center Southborough, Massachusetts
		Dr. S. S. Kalter Virus Reference Laboratory, Inc. San Antonio, Texas
		Dr. Donald G. Lindburg Center for Reproduction of Endangered Species San Diego, California
		Dr. Roger Mast Conservation International Washington D.C.
		Dr. Thomas E. Nolan Merck Sharpe & Dohme Laboratories West Point, Pennsylvania
		Dr. Joyce Siriani American Society of Primatology SUNY Buffalo, New York
		Dr. Thomas Wolfle National Academe fo Science Washington, D.C.
		Dr. G. L. Van Hoosier ICLAS University of Washington Seattle, Washington
VENEZUELA VENEZUELA	Representative Representante	Dra. Roberta Bodina Directiva Sociedad Venezolana de Primatología Caracas

GOVERNMENTS GOBIERNOS	ROLE INTERVENCION	NAME NOMBRE
Battelle Pacific Northwest Laboratories Richland, Washington	Host Anfitron	Dr. Janet Baer Life Sciences Center
		Dr. Edward Liebow Human Affairs Research Center
		Mr. Ronald C. Liikala Strategic Planning
		MS Tonya Quackenbush, Life Sciences Center
		Dr. Richand E. Weller Life Sciences Center
		Dr. William R. Wiley Director
Pan American Health Organization/ World Health Organización	Secretariat Secretariá	Dr. Primo Arámbulo III Program Coordinator Veterinary Public Health Washington D.C.
		Dr. Albino Belotto Asesor Salud Publica Veterinania La Paz, Bolivia
		Dr. Filomeno Encarnación Proyecto Peruano de Primatología Iquitos, Perú
		Dr. Jorge Escalante Consultant, Veterinary Public Health Washington D.C.
		Dr. Luis Moya Proyecto Peruano de Primatología Iquitos, Perú
		Dr. Jaime Moro Proyecto Peruano de Primatología Iquitos, Perú
		Dr. Hugo Samamé Consultant, Primatology Iquitos, Perú

PROGRAM OF SESSIONS / PROGRAMA DE SESIONES

Monday 29 October / Lunes 29 de Octubre

FIRST PLENARY SESSION / PRIMERA SESION PLENARIA

8:00 a.m.	Registration / Inscripcion	
	Item 1:	Restricted to official delegates of the member countries of the Committee.
	Tema 1:	Reunion restringida de delegados oficiales de los paises miembros del Comite.

INAGURAL SESSION / SESION INAUGURAL

10:00 a.m.	Opening Ceremony / Ceremonia de Inauguracion -
10:30 a.m.	Break / Receso

SECOND PLENARY SESSION: CONSERVATION, MANAGEMENT, AND HEALTH / SEGUNDA SESION PLENARIA: CONSERVACION, MANEJO Y SALUD

11:00 a.m.	Item 2:	Current Situation and Conservation of Neotropical Primates in their Habitat (Document CORP I/4)
	Tema 2:	Estado Actual y Conservacion de los Primates Neotropicales en su Propio Habitat (Documento CORP I/4) Dr. Roderic Mast
11:20 a.m.	Discussion / Discusion	
11:50 a.m.	Item 3:	Status of Natural Primate Populations and the Benefits of Periodic Cropping (Document CORP I/5)
	Tema 3	El Estado de las Poblaciones Naturales de Primates y los Beneficios de la Cosecha Sostenida (Documento CORP I/5) Dr. Luis Moya
12:10 p.m.	Discussion / Discusion	
12:40 p.m.	Break / Receso	
2:30 p.m	Item 4 :	Environmental Factors Affecting the Reproduction of Primates in Special Habitats (Document CORP I/6)
	Tema 4 :	Los Factores Ambientales para la Reproduccion de Primates en Habitat Especiales (Documento CORP I/6) Dr. Thomas Wolfle

2:50 p.m.	Discussion / Discusion	
3:20 p.m.	Item 5:	The Conservation of Nonhuman Primates and its Importance to Public Health (Document CORP I/7)
	Tema 5:	La Conservacion de los Primates No Humanos y su Importancia en la Salud Publica (Documento CORP I/7) Dr. Robert Whitney/Dr. Sephen Potkay
3:40 p.m.	Discussion / Discusion	
4:10 p.m.	Break / Receso	
4:25 p.m.	Item 6:	Program of Technical Cooperation of the Pan American Health Organization for the Conservation, Reproduction, and Utilization of Nonhuman Primates (Document CORP I/8)
	Tema 6:	El Programa de Cooperacion tecnica de la Organizacion Panamericana de la Salud para la conservacion, reproduccion y utilizacion de los primates no humanos. (Documento CORP I/8) Dr. Primo Arambulo III
4:45 p.m.	Discussion / Discusion	
5:15 p.m.	Break / Receso	

Tuesday 30 October / Martes 30 de Octubre

THIRD PLENARY SESSION: POLITICS AND ECONOMICS /
TERCERA SESION PLENARIA: POLITICA Y ECONOMIA

9:15 a.m.	Item 7:	Brazil's Fauna Management (Document CORP I/9)
	Tema 7:	Manejo de Fauna en Brasil (Documento CORP I/9) Dr. Jordan Paulo Walluer
9:35 a.m.	Discussion / Discusion	
10:05 a.m.	Item 8:	Managing Non-human Primates as Renewable Resources: The Economics of Conservation (Document CORP I/10)
	Tema 8:	Manejo de los Primates No Humanos como Recurso Renovable: La Economia de la Conservacion (Documento CORP I/10) Dr. Theodore Panayotou/ Dr. Primo Arambulo III

10:25 a.m.	Discussion / Discusion	
10:55 a.m.	Break / Receso	
11:10 a.m.	Item 9:	Integrated Development of Conservation Units and the Role of National Primatology Programs (Document CORP I/11)
	Tema 9:	El Desarrollo Integrado de las Unidades de Conservacion y el Rol de los Programas Nacionales de Primatologia (Documento CORP I/11) Dr. Vicente Rodriagez
11:30 a.m.	Discussion / Discusion	
12:00 p.m.	Break / Receso	
2:00 p.m.	Item 10:	Multisectoral and Intercountry Cooperation for the Development of National Primatology Programs (Document CORP I/12)
	Tema 10:	La Cooperacion Multisectorial e Interpaises para el Desarrollo de Programas Nacionales de Primatologia (Documento CORP I/12) Mr. Manuel Uceda
2:20 p.m.	Discussion / Discusion	
2:50 p.m.	Item 11:	The Mobilization of Resources for the Conservation of Neotropical Ecosystems (Document CORP I/13)
	Tema 11:	La Movilizacion de Recursos para Conservacion del Ecosistema (Documento CORP I/13) Dr. Edward Liebow
3:10 p.m.	Discussion / Discusion	
3:40 p.m.	Item 12:	The Community Participation in the Protection of Neotropical Primates Natural Habitat and the Policy of Natural Resources in Bolivia (Document CORP I/14)
	Tema 12:	La Participacion de las Comunidades en la Proteccion del Habitat Natural de los Primates Neotropicales y la Politica de Recursos Naturales en Bolivia (Documento CORP I/14) Dr. Hernan Zeballos
4:00 p.m.	Discussion / Discusion	

4:30 p.m.	Item 13:	Regional Plan of Action for Primatology in the Americas (Document CORP I/15)
	Tema 13:	El Plan de Accion Regional de Primatologia para las Americas. (Documento CORP I/15) Dr. Filomeno Encarnacion/ Dr. Huao Samame
5:00 p.m.	Discussion / Discusion	

Wednesday 31 October / Miercoles 31 de Octubre

CLOSING SESSION / SESION DE CLAUSURA

9:00 a.m.	Reading of the Recommendations and signature of the Final Report Lectura de las recomendaciones y firma del informe final.
12:00 p.m.	Closing Ceremony / Clausura

PROPOSED CONSTITUTION OF THE REGIONAL PRIMATOLOGY COMMITTEE FOR THE AMERICAS

Article I
Background

The Declaration on Technical Cooperation among the Countries of the Americas in the Conservation and Utilization of Nonhuman Primates, known as the Declaration of Iquitos, signed by the 50 participants from America and Europe at the Workshop on Controlled Breeding of Primates in their Natural Habitat, held in Iquitos in November 1986, recommends the creation of a Regional Committee.

This declaration was submitted to the V Inter-American Meeting, at the Ministerial Level, on Animal Health held in April 1987, which unanimously approved Resolution VIII, which creates the Regional Primatology Committee and names the Pan American Health Organization (PAHO) as Secretariat ex officio (hereafter Secretariat).

Article II
Purposes

The Regional Primatology Committee (CORP) shall have as its purposes ensuring the maintenance of the political will of the Member Countries of PAHO to safeguard the life of the species of neotropical primates in their natural habitat, and strengthening and supporting technical cooperation among the countries with native populations of primates, within a regional plan of action that will guarantee the preservation of the natural ecosystem, and will seek to improve living conditions for humans.

Article III
Headquarters

The headquarters of CORP shall be at the Pan American Health Organization, in Washington, D. C.

Article IV
Organization

CORP shall consist of an Assembly and a Governing Board, which includes the Secretariat.

1. The Assembly

The Assembly shall consist of the official representatives of those Member Countries of PAHO that have within their territories native populations of nonhuman primates and which agree to conduct official programs for the conservation of the species and their habitat.

Also members are the official representatives of the countries that contribute financially to the conservation of primates. All members shall have a right to voice and vote.

The Chairman of the Assembly shall be Chairman of the Governing Board, and the Secretariat shall be the Rapporteur.

The Assembly may include observer members who represent the scientific and economic areas of national and international importance within the framework of conservation of natural resources and neotropical ecosystems. They shall have a voice, with consent of the Chairman, but they shall not have a vote.

2. The Governing Board

The Governing Board consists of three member representatives from the Assembly, and the Secretariat ex officio. There shall be a Chairman, two members representing regions of the Americas that the Chairman does not represent, and the Secretariat. The Chairman shall preside over the Assembly and the Governing Board.

3. The Secretariat

In accordance with Resolution VIII of RIMSA V, the Pan American Health Organization shall act as the Secretariat ex officio of CORP. By delegation of the Director of PAHO, the Program for Veterinary Public Health shall act as the Secretariat.

4. The Meetings

CORP shall hold regular and special meetings. Regular meetings shall be held once a year, and special meetings when circumstances so require or when requested by a majority of the members.

The meetings of the Committee shall be convened by the Secretariat in the name of the Chairman, to be held at the headquarters of the Committee, or in the place agreed to by the members.

5. The Commissions

CORP shall have the power to establish special, temporary, or permanent Commissions on specific topics related to primatology in the Americas, and such others as the Committee shall determine.

Members of the commissions shall be specialists of recognized prestige, who may be nationals of the member countries of the Committee, or of other countries.

Article V
Admission and/or incorporation of new members

The Secretariat shall communicate to the Committee all requests for membership. The admission of other countries shall be decided by the Committee, and shall be effective from the date of approval.

Article VI
Withdrawal

The Secretariat shall communicate to the Committee all notices of withdrawal from the Committee. Any Member Country may withdraw from CORP by a written notice given not less than six months prior to the next regular meeting.

Article VII
Amendment of the Constitution

This Constitution may be amended in a regular meeting at the request of one of the members of CORP; such amendment requires the approval of the majority of the members.

PROPOSED REGULATIONS OF THE REGIONAL PRIMATOLOGY COMMITTEE FOR THE AMERICAS (CORP)

Article I. Background

Resolution VIII of the V Inter-American Meeting, at the Ministerial Level, on Animal Health (RIMSA V), held in Washington, D.C., from 27 to 30 April 1987, created the Regional Primatology Committee for the Americas (CORP). It also provided that the Pan American Health Organization (PAHO) shall act as Secretariat ex officio of the Committee.

Article II. Purposes

CORP shall have as its purposes ensuring the maintenance of the political will of the Member Countries to safeguard the life of the species of neotropical primates in their natural habitat, and strengthening and supporting technical cooperation among the countries with native populations of primates.

Article III. Objectives

To achieve these purposes, the Committee shall have the following objectives:

1. To prepare and approve a regional plan of action to serve as a frame of reference for national plans.

2. To develop and promote comprehensive programs in those ecosystems with natural populations of primates, aimed at improving the living conditions of the human inhabitants, and preserving the wild fauna.

3. To prepare technical outlines and recommendations for proper evaluation of the ecosystems.

4. To promote research programs, to provide advisory services, and to exchange the results obtained.

5. To seek understanding among the countries so that plans for the building of infrastructure, colonization, and urban expansion do not interfere with efforts being made to preserve the environment.

6. To promote the mobilization of national resources for conservation of the natural habitat of primates.

7. To promote the installation of breeding colonies that meet the control standards in order to satisfy the local needs and the international demands

Article IV. Members

The following countries shall be members of the Committee:

Argentina, Barbados, Bolivia, Brazil, Canada, Colombia, Mexico, Paraguay, Peru, United States of America, and Venezuela and the countries of Central America, who will be represented by the corresponding state organizaton or whom ever they designate, duly accredited as an official delegate.

The official delegate will represent its Government who will authorize him to attend all the meetings of the Committee convened by the Secretariat.

Article V. Observers

PAHO Member or Participating Governments that are not members of the Committee may be present at the meetings as observers.

The Committee may invite as observers representatives of other countries and of nongovernmental international organizations involved with the conservation of primates.

Observers may participate in discussions without vote, and present documents for consideration by the Committee.

Article VI. Secretariat

PAHO shall act as Secretariat of the Committee, and shall serve as Rapporteur in the regular and special meetings.

The functions of the Secretariat shall be to convene regular and special meetings, to receive and distribute documents, reports, recommendations, and resolutions of the Committee, to prepare minutes of the meetings, and to carry out such other tasks as may be required by the Committee.

The Secretariat shall be responsible for the follow-up of the resolutions approved by the Committee, and shall report the results of such follow-up in the next regular meeting.

Article VII. Headquarters

The Committee shall have its headquarters at the offices of PAHO in Washington, D.C.

Article VIII. Organization

The Committee shall be presided over by the representative of one of the Member Countries, to be designated following the alphabetical order of the countries. The chairmanship shall be for a period of one year from the date of the regular meeting in which the designation was made.

If the representative of the country designated for the chairmanship is absent, the chairmanship shall be exercised by the representative of the country to be designated for the following period, i.e., the country next in alphabetical order.

Article IX. Meetings

The Committee shall hold two types of meetings:

(a) regular, which shall be held once a year, and

(b) special, which shall be convened when circumstance so require, or when a majority of the members so request.

Meetings of the Committee shall be convened by the Secretariat in the name of the Chairman, and shall be held at the headquarters of the Committee, or in a place agreed upon by its members.

The provisional agenda and schedule of sessions shall be prepared by the Secretariat and sent out together with the notice of convocation of the meeting. Members of the Committee may propose additional items for the agenda.

Decisions of the Committee shall be by majority vote, and shall be in the form of recommendations.

Reports of the meetings of the Committee and the pertinent recommendations shall be forwarded by the Secretariat to the respective Ministers of Agriculture of the Member Countries for their consideration and compliance.

The Secretariat shall also send the reports of the meetings of the Committee to the countries and international organizations that have accredited observers, and to other countries and international organizations that so request.

The official representatives of the countries are obligated to make arrangements at the national level to implement the recommendations approved, and they shall report on the results at the next regular meeting.

Article X. Governing Board

The Committee shall have a Governing Board consisting of a Chairman, two members representing regions of the Americas that the Chairman does not represent, and the Secretariat.

The Governing Board shall be elected annually in the regular meeting of the Committee, and shall represent it when so delegated by the Committee.

The Chairman shall preside over the Governing Board and the regular and special meetings during his designated period.

Article XI. Decisions and Resolutions

Each member of the Committee shall have the right to a vote. The decisions of the Committee shall be adopted preferably by consensus or, lacking this, by a majority of the votes cast, and they shall be drafted as recommendations.

Article XII. Commissions

The Committee may establish special, temporary, or permanent commissions to carry out studies and report on matters relating to the conservation and utilization of nonhuman primates, as well as on other matters that the Committee may specifically direct.

Members of the commissions shall be specialists of recognized prestige in the specific area involved, and may be nationals of the Member Countries of the Committee, or of other countries.

The commissions shall be convened by the Secretariat on the date and in the place most suitable for the purposes of each commission.

Article XIII. Admission

The admission of other countries shall be by decision of the Committee, and shall be effective from the date on which the respective request for membership is approved. The Secretary shall communicate the receipt of each request for membership to the Member Countries of the Committee.

Article XIV. Withdrawal

Any Member Country may withdraw from the Committee by addressing a written notice to the Secretariat, presented not less than six months prior to the date of the next regular meeting. The secretariat shall communicate to the Member Countries the receipt of each notice of withdrawal.

Article XV. Financial

The Member Countries shall agree to make annual contributions to finance the work of the Committee or specifically approved projects or research, which shall be carried out in coordination with the Secretariat.

Funds may be allocated to finance expenditures involved in the organization of the meetings of the Committee and the commissions as provided for in the Constitution, and in such other projects as may be approved by the Committee.

PAHO shall contribute the necessary resources for Secretariat services.

Article XVI. Amendments to the Regulations

These Regulations may be amended in regular meetings at the request of a Member Country of the Committee, with the approval of a majority of the members. The amendments shall be sent to the respective countries, as approved, and shall become an integral part of these Regulations.

Article XVII. Termination

The Committee may cease to function through a resolution approved by two-thirds of its members and ratified by RIMSA.

Article XVIII. Other Matters

All other matters not provided for in these Regulations shall be resolved directly by the members of the Committee in any regular or special meeting.

ESTATUTOS DEL COMITÉ REGIONAL DE PRIMATOLOGÍA PARA LAS AMÉRICAS

Artículo I
De los antecedentes

La declaración sobre "La Cooperacion técnica entre los paises de las Américas en la conservacion y utilizacion de los primates no humanos", conocida como la Declaración de Iquitos, firmada por los 50 participantes, de América y Europa, al "Taller sobre Crianza controlada de primates en su propio habitat" llevada a cabo en Iquitos en Noviembre de 1986 recomienda la creación de un Comité Regional.

Esta declaracion fue presentada a la V Reunión Inter Americana de Salud Animal a Nivel Ministerial realizada en Abril de 1987, la que por unanimidad aprobó la Resolución VIII por la cual se crea el Comité Regional de Primatología, encargando a la Organización Panamericana de la Salud (OPS) para que actúe como Secretaria ex-officio.

Artículo II
De los oropositos

El Comité Regional de Primatología (CORP) tiene como propósito el asegurar el mantenimiento de la voluntad politica de los países miembros de la OPS para salvaguardar la vida de las especies de primates neotropicales en su habitat natural, fortalecer y apoyar la cooperación técnica entre los países con poblaciones de primates, dentro de un plan de acción regional que garantice la conservación del ecosistema natural, que contribuya a mejorar la condición de vida humana.

Artículo III
De la Sede

El CORP tiene su sede en las instalaciones de la Organización Panamericana de la Salud en Washingon, D.C. Estados Unidos de America.

Artículo IV
De la Oraanizacion

El CORP esta constituido por una Asamblea y una Junta Directiva que incluye a la Secretaria ex-officio.

1. De La Asamblea

Esta integrada por los representantes oficiales de los países miembros de la OPS que en sus territorios tienen poblaciones nativas de primates no humanos y que acuerdan conducir programas oficiales para la conservación de las especies y de su habitat. También son miembros los representantes oficiales de los países que contribuyen financieramente para la conservación de los primates. Todos tienen derecho a voz y voto.

El Presidente de la Asamblea será el Presidente de la Junta Directiva y el Relator será la Secretaria ex-officio.

La Asamblea del CORP puede incluir miembros observadores que representan a las áreas científicas y economicas de relevancia nacional e internacional en el marco de la conservación de los recursos naturales y los ecosistemas neotropicales. Tienen voz por anuencia de la Presidencia, pero no tienen voto.

2. De La Junta Directiva

Esta conformada por cuatro representantes miembros de la Asamblea y la Secretaria ex-officio. Los cargos son los siguientes: Un Presidente, un vicepresidente, dos vocales representando las regiones de las Américas que el Presidente no representa y la Secretaría ex-officio. El Presidente dirige la Asamblea y la Junta Directiva.

3. De la Secretaría

En concordancia con la Resolución VIII del RIMSA V la Organización Panamericana de la Salud actuará como Secretaría ex-officio del CORP. Por delegación de funciones del Director de la OPS, el Programa de Salud Pública Veterinaria fungirá como la Secretaría ex-officio.

4. De las Reuníones

El CORP celebrará Reuniones ordinarias y extraordinarias. Las ordinarias se llevarán a cabo una vez al año y las extraordinarias cuando las circunstancias 10 exijan o cuando 10 soliciten la mayoria de los miembros.

Las reuniones del Comité serán convocadas por la Secretaría ex-officio en nombre del Presidente, debiendo celebrarse en la Sede del Comité o en el lugar que acuerden sus miembros.

5. De las Comisiones

El CORP tiene la facultad de constituir Comisiones especiales temporales o permanentes sobre determinados temas relacionados a la primatología en las Américas y otros que indique el Comité.

Los integrantes de las Comisiones son especialistas de prestigio reconocido, pudiendo ser nacionales de los países miembros del Comité o de otros.

Artículo V
De la admisión y/o incorporación de nuevos miembros

La Secretaría ex-officio comunicará toda solicitud de ingreso al Comité. La admisión de otros países queda a consideración del Comité y tendrá efecto desde la fecha de aprobación.

Artículo VI
Del Retiro

La Secretaría ex-officio comunicara toda solicitud de retiro al Comité. Cualquier pais miembro puede retirarse del CORP mediante comunicación escrita con una antelación no menor de seis meses a la siguiente reunión ordinaria.

Artículo VII
De las Modificaeión del Estatuto

El presente estatuto puede ser revisado en reunión ordinaria a solicitud de uno de los miembros del CORP y modificado con la aprobación de la mayoría.

REGLAMENTO DEL COMITÉ REGIONAL DE PRIMATOLOGÍA PARA LAS AMÉRICAS (CORP)

Artículo I De los Antecedentes

La Resolucion XIII de la V Reunión Interamericana a Nivel Ministerial de Salud Animal (RIMSA V), realizada en Washington, D.C., del 27 al 30 de abril de 1987, crea el Comité Regional de Primatología para las Americas (CORP). Se establece, además, que la OPS, actue como Secretaría ex-officio del Comité.

Artículo II De los Propósitos

El (CORP) tiene como propósito el asegurar el mantenimiento de la voluntad política de los países miembros para salvaguardar la vida de las especies de primates especialmente las neotropicales en su habitat natural, fortalecer y apoyar la cooperación técnica entre los países con poblaciones nativas de primates.

Artículo III De los Objetivos

Para alcanzar el propósito enunciado, el comité tendrá los siguientes objetivos:

1. Elaborar y acrobar un plan de acción Regional como marco de referencia para planes nacionales, adecuado a las necesidades de cada país.

2. Desarrollar y promover programas integrales en los ecosistemas con poblaciones naturales de primates tendientes a mejorar la condición de vida del poblador humano, y a la conservación de la fauna silvestre.

3. Elaborar lineamentos técnicos y recomendaciones para la adecuada evaluación de los ecosistemas.

4. Promover programas de investigación, prestar su asesoramiento, el intercambio de los resultados obtenidos, y capacitación de recursos humanos.

5. Procurar el entendimiento entre los países para que los planes de construcciones de infraestructura, colonizacion, expansión urbana, no afecten los esfuerzos desplegados para la conservación del medio ambiente.

6. Promover la movilización de recursos nacionales para la conservación de los habitat naturales de los primates.

7. Promover la instalación de criaderos en condiciones controladas para satisfacer necesidades locales y demanda internacional.

Artículo IV De los Miembros Natos

Serán miembros del Comité los siguientes países: Argentina, Barbados, Bolivia, Brasil, Canada, Colombia, Mexico, Peru, Estados Unidos de America, Paraguay, Venezuela, y los países de Centroamérica los que estarán representados por el organismo estatal correspondiente o por quien ellos designen, debidamente acreditado como delegado oficial.

El delegado oficial representará a su gobierno, el cual lo autorizará a asistir a las reuniones del Comité que sean convocadas por la Secretaría.

Artículo V De los Miembros Observadores

Todo Gobierno Miembro ó Gobierno Participante de la Organización Panamericana de la Salud que no forme parte del Comité podrá hacerse representar en la reuniones por observadores.

El Comité podrá invitar como observadores a representantes de otros países y de organismos internacionales no gubernamentales vinculados con la conservación de los primates.

Los observadores podrán participar de los debates sin derecho a voto y presentar documentos a consideración de la Comision.

Artículo VI De la Secretaría

La Organizacion Panamericana de la Salud, actuará como Secretaría ex-officio del Comité y cumplirá las funciones de Relator en la reuniones ordinarias y extraordinarias.

Seran funciones de la Secretaría convocar a las reuniones ordinarias y extraordinarias, recibir y distribuir los documentos, informes, recomendaciones y resoluciones de la Comisión, preparar las actas de las reuniones y realizar otras tareas a requerimiento del Comité.

La Secretaría será responsable del seguimiento de las resoluciones aprobadas por el Comité, debiendo dar cuenta de los resultados de ese seguimiento en la siguiente reunión ordinaria.